ASTROPHYSICS
AND
THE EVOLUTION
OF THE UNIVERSE

ASTROPHYSICS
AND
THE EVOLUTION
OF THE UNIVERSE

Leonard S Kisslinger
Carnegie Mellon University, USA

World Scientific

NEW JERSEY · LONDON · SINGAPORE · BEIJING · SHANGHAI · HONG KONG · TAIPEI · CHENNAI

Published by

World Scientific Publishing Co. Pte. Ltd.
5 Toh Tuck Link, Singapore 596224
USA office: 27 Warren Street, Suite 401-402, Hackensack, NJ 07601
UK office: 57 Shelton Street, Covent Garden, London WC2H 9HE

Library of Congress Cataloging-in-Publication Data
Kisslinger, Leonard S., author.
    Astrophysics and the evolution of the universe / Leonard S Kisslinger, Carnegie Mellon University, USA.
    pages cm
    Includes bibliographical references and index.
    ISBN 978-9814520904 (hardcover : alk. paper)
    1. Astrophysics--Textbooks. I. Title.
    QB461.K565 2014
    523.01--dc23
                        ,
                            2013049520

British Library Cataloguing-in-Publication Data
A catalogue record for this book is available from the British Library.

Typeset by Stallion Press
Email: enquiries@stallionpress.com

Printed in Singapore

Preface

This book is written for college students who might not have taken any college science or mathematics courses. It is intended to inform students of any major about the exciting and somewhat mysterious developments in astrophysics in recent decades. It explains how our universe evolved from the very earliest time to the present using very advanced theories presented in way that a college student with high school mathematics can understand.

Objectives

1. To discuss the basic physics and astrophysics concepts needed to understand the content and structure of the universe.
2. To present the Special and General Theories of Relativity needed to treat the evolution of the universe.
3. To introduce important events like inflation and cosmological phase transition that occurred at well-known times in a manner that can be understood and appreciated by the college students who take this course.

Contents

Chapter 1. Physics Concepts Needed for Astrophysics. The basic concepts of position, velocity, acceleration, momentum, energy, and temperature as a form of energy are defined and discussed. Force and Newton's Law of Motion are discussed, especially the force of gravity, which is the basic force needed to derive the evolution of the universe. Dark Energy is briefly introduced.

Chapter 2. Forces and Particles. In this chapter the Standard Model elementary particles are discussed. The fermions are leptons and quarks, and the vector bosons are the quanta of the electromagnetic, strong and weak interactions. The Higgs, a scalar boson, is the other Standard Model particle. Composite physical objects such as atoms, atomic nuclei, baryons and mesons are also discussed. Matter beyond the Standard Model, which will be shown to dominate the matter in the universe in later chapters, is defined.

Chapter 3. Hubble's Law: Expansion of the Universe. First the Doppler shift and the redshift of light from a source moving away from an observer are defined and discussed. Then Hubble's Law, derived from measuring the Doppler shift of light from galaxies, is reviewed. The expansion of the universe is discussed.

Chapter 4. Stars, Galaxies, Etc. The concept of centripetal acceleration is reviewed to understand how the earth moves around the sun, and the more complex concepts related to our solar system being a disk with rotating planets. The sun as a fusion furnace is discussed, as well as the first observation of Dark Matter by measuring the rotational speed of our Milky Way galaxy. Finally the creation of pulsars and black holes by the gravitational collapse of a massive star is reviewed.

Chapter 5. Neutrino Oscillations, Symmetries, and Pulsar Kicks. An important property of the three Standard Model neutrinos changing into each other, called neutrino oscillations, is explained. Parity, charge conjugation, and time reversal symmetries are defined, and how neutrino oscillations can be used to test these symmetries is discussed. Finally, possible cause of the rapid velocities of pulsars, called pulsar kicks, by neutrino emission is discussed.

Chapter 6. Einstein's Special and General Theories of Relativity. First, the main postulate in Einstein's Special Theory of Relativity, that the speed of light in vacuum measured in any inertial frame is equal to c, and the resulting length contraction and time dilation are discussed. Einstein's formula for the addition of velocities, derived from the Lorentz transformation, is shown to be consistent with his postulate. Relativistic momentum is defined and the law of motion is shown to be the same as Newton's except for different definitions of momentum. Einstein's General Theory of Relativity is discussed using tensors, which are defined. The much simpler equations of General Relativity used to find the evolution of the universe are defined and discussed in Chapter 7.

Chapter 7. Radius and Temperature of the Universe from the General Theory of Relativity. First tests of the Special Theory of Relativity via length contraction and time dilation, as well as energy and momentum conservation to test the relativistic definitions of these quantities are discussed. Next tests of Einstein's General Theory of Relativity via the gravitational Doppler shift and gravitational deflection of light are discussed. The Schwarzschild radius for black holes is given. Then the main topic, Friedmann's equation, a much simpler form of General Relativity than Einstein's equations using tensors, is derived, from which the $R(t)$ and $T(t)$, the radius and temperature of the Universe as a function of time are derived. Then Friedmann's equation with a cosmological constant, representing Dark Energy, are used to derive inflation, which produced the universe's almost homogeneous nature. Then gravitational radiation and gravitational quantum theory, with the graviton the quantum of the gravitational field are briefly discussed.

Chapter 8. Cosmic Microwave Background Radiation (CMBR). This chapter started with a derivation of the quantum Planck spectrum, from which the temperature of a black box can be determined, and the temperature of our universe has been measured. The main topic is T–T correlations in the Cosmic Microwave Background Radiation (CMBR), arising from acoustic oscillations, from which the amount of visible and Dark Matter, as well as Dark Energy and other properties of the universe, such as the age of the universe, have been determined.

Chapter 9. Electroweak Phase Transition (EWPT). The chapter starts with a brief discussion of classical phase transitions, with the important concept of latent heat defined. After introducing the basic concepts in quantum theory (states, operators, and expectation values of operators), quantum mechanical phase transitions and latent heat are defined. Electroweak theory is introduced and the EWPT is discussed, with the expectation value of the Higgs quantum field being 0 before and the Higgs mass after the EWPT.

The creation of magnetic fields during the EWPT and their production of gravity waves are reviewed. Einstein's tensorial form for General Relativity and the derivation of Friedmann's equation is carried out in order to introduce gravitational quantum theory, with the graviton, and gravitational wave creation during the EWPT.

Chapter 10. Quantum Chromodynamic Phase Transition (QCDPT). The QCDPT was a transition at about 10^{-5} s from a quark–gluon plasma (QGP) universe to one with protons and neutrons. The quark condensate $\langle \bar{q} q \rangle$, with q the quark field, the latent heat for the QCDPT, is defined and its value given. The magnetic field B_W produced by QCDPT bubbles is derived and B–B correlations in the CMBR calculated. The evolution of this magnetic field to the present time indicates that it could be the seeds for present galactic magnetic fields, solving an astrophysical puzzle. It is shown that the magnitude of the quark condensate is consistent with its being the source of Dark Energy. Finally, the detection of a QGP produced by relativistic heavy ion collisions is discussed.

Solutions to Problems. The solutions to the problem at the end of the ten chapters are given to help the student understand the concepts and results, many of which are quite advanced.

Appendix: Vector Calculus and Maxwell's Equations of Electricity and Magnetism

Contents

Chapter 1

Physics Concepts Needed for Astrophysics

In this first chapter, after a brief overview of some basic physics concepts we review the concepts of position, velocity, and acceleration as vectors, needed for the treatment of topics such as the size and rate of expansion of the universe. We then discuss force, Newton's Law of Motion, and the force of gravity, and review the concept of energy and temperature as a form of energy. As we discuss in later chapters, knowing the temperature of the universe at different times is essential to understand the evolution of the universe. Finally we briefly introduce Dark Energy.

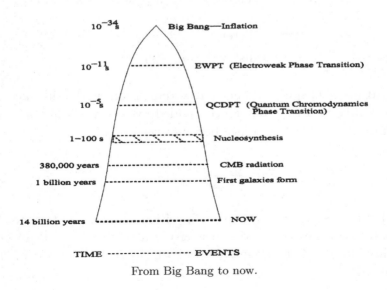

From Big Bang to now.

1

The figure shows the evolution of the universe from the very early time, when inflation took place, through cosmological phase transitions EWPT (particles got masses) and QCDPT (neutrons and protons form), emitted the Cosmic Microwave Background Radiation (CMBR) from which we learn the constituents of the universe, to the present time with stars and galaxies. We will explore all of these events in this book.

1. Overview of Some Basic Physics Concepts

The basic concepts reviewed in this chapter are treated in greater detail in some textbooks, such as Blatt [1], Halliday and Resnick, or some more recent texts.

1.1. *Time, Position, Distance, Velocity, Acceleration, Force, Newton's Law of Motion, Energy, Temperature*

1.1.1. *Time $= t$, Position $= r$, Distance, Velocity*

Calling time $= t$ and position $= r$, at time $= t_1$ an object is at position r_1, and at time t_2 at position r_2. Let us define the difference in time as Δt, and the difference between positions as distance:

$$\text{time difference} \equiv \Delta t = t_2 - t_1$$
$$\text{distance} = r_2 - r_1. \tag{1}$$

Velocity $= v$, speed in one spatial dimension, is the distance an object moves in a time difference divided by the time difference, in the limit as the time difference becomes very small:

$$v = \left.\frac{(r_2 - r_1)}{\Delta t}\right|_{\Delta t \to 0}. \tag{2}$$

That is, v is the rate of change of position with respect to time. As we shall see in the next section, velocity is a vector, with magnitude and direction, and the quantity $v =$ speed is the magnitude of the velocity.

1.1.2. *Acceleration, Force, and Newton's Law of Motion*

If v_1, v_2 are speeds at t_1, t_2, then acceleration $= a$ is the rate of change of speed with respect to time:

$$a = \frac{(v_2 - v_1)}{\Delta t}\bigg|_{\Delta t \to 0}. \tag{3}$$

Force $= F$ causes an object to move. That is, it changes its speed. From the definitions above, force causes acceleration.

NEWTON'S LAW OF MOTION: if a force F acts on an object with mass m, then m has acceleration $= a$, which is

$$a = F/m. \tag{4}$$

Newton's Law of Motion, which we study in detail in Section 3, is one of the most important principles in science.

1.1.3. *Work and Energy*

Work: If a force is exerted on an object and it moves a distance $= d$, then the force does work $= W =$ force times distance. If work is exerted on a mass m, m gets energy, E, equal to the work:

$$W = E.$$

The types of energy are:

KINETIC ENERGY $=$ energy of motion.
POTENTIAL ENERGY $=$ stored energy.
HEAT ENERGY $=$ energy of a substance due to its components, such as atoms, moving. A type of internal kinetic energy.

TEMPERATURE $= T$ of an object is a measurement of its heat energy, with heat energy $= kT$, where k is Boltzman's constant. As we shall see in later chapters, T is a very important property when we explore the evolution of the universe.

We shall study energy in detail later in this chapter.

1.2. *Units of Time, Distance, Velocity, Acceleration, Force, Energy, Temperature*

1.2.1. *Powers of Ten*

Before we discuss the units, we need to define the powers of ten.

$$10^n \equiv 1 \text{ followed by } n \text{ zeros}$$
$$10^1 = 10; \quad 10^2 = 100;$$
$$10^6 = 1{,}000{,}000 = \text{one million};$$
$$10^9 = 1{,}000{,}000{,}000 = \text{one billion}. \qquad (5)$$

1.2.2. *Units of Time, Distance, Velocity, and Acceleration*

The unit of time is second = sec = s

$$1 \text{ minute} = 60\,\text{s}$$
$$1 \text{ hour} = 60 \text{ minutes} = 60 \times 60\,\text{s} = 3{,}600\,\text{s} = 3.6 \times 10^3\,\text{s}$$
$$1 \text{ year} = 365 \text{ days} = 365 \times 24 \times 3{,}600\,\text{s} = 31{,}536{,}000\,\text{s}$$
$$= 3.1536 \times 10^7\,\text{s} \simeq 3.15 \times 10^7\,\text{s}$$

The unit of distance or position is m = meter. One also uses cm = centimeter.

$$1\,\text{m} = 100\,\text{cm}$$
$$1 \text{ kilometer} = 1\,\text{km} = 1000\,\text{m} = 10{,}000\,\text{cm}, \quad \text{or} \quad 1\,\text{km} = 10^5\,\text{cm}$$

Astronomical units of distance:

$$1 \text{ lightyear} = 1 \text{ ly} = \text{distance light travels in 1 year}$$
$$1 \text{ ly} = 300{,}000\,\text{km/s} \times 3.15 \times 10^7\,\text{s} = 9.46 \times 10^{15}\,\text{m}$$
$$1 \text{ pc} = 1 \text{ parsec} = 3.25 \text{ lightyears} = 3 \times 10^{18}\,\text{cm}$$

The basic unit of velocity is km/s. An important velocity (or speed) is the speed of light:

$$\text{Speed of light} = c = 300{,}000\,\text{km/s}$$

The basic unit of acceleration is m/s^2. For example, if you drop an object near the surface of the earth, it will feel an acceleration g,

$$\text{acceleration of gravity} = g = 9.80\,\text{m/s}^2, \qquad (6)$$

which we prove when we discuss the force of gravity.

1.2.3. *Units of Force, Energy, and Temperature*

The standard unit of force = newton = N.

Remembering work = force times distance, the unit of work = Nm = joule = J. Since work produces energy, unit of energy, kinetic or potential is

$$\text{joule} = J = 1.0 \, \text{kg m}^2/\text{s}^2.$$

In our study of astrophysics we most often use the unit of electron volt = eV for energy, which we describe later in this chapter. It is related to the force of electricity, which we discuss in Chapter 2.

$$1 \, \text{MeV} = 1{,}000{,}000 \, \text{eV} = 10^6 \, \text{eV}; \quad 1 \, \text{GeV} = 10^9 \, \text{eV}$$

The standard unit of temperature is °C (Celcius). $0°C = 273K$, where K is kelvin. One °C has the same magnitude as one K. 0K is absolute zero temperature. As we shall see, the universe has a temperature of just a few degrees above absolute zero. Also $1°C = 1.8°F$, with F being Fahrenheit, which is used in the US but not in most other countries.

Important examples of temperature are phase transitions, such as water, a liquid, being heated until it becomes steam, a gas. We shall study this later in the chapter.

1.2.4. *Mass Energy*

A famous relation from Einstein's Special Theory of Relativity is $E = mc^2$ which means that a mass m has a mass energy mc^2. Thus an object with mass m can decay into particles with smaller masses, and they will have kinetic energy as well as mass energy.

$$\text{electron mass energy: } 1 \, m_e c^2 = 0.511 \, \text{MeV} = 5.11 \times 10^5 \, \text{eV}$$
$$\text{proton mass energy: } 1 \, m_p c^2 \simeq 1.0 \, \text{GeV} = 10^9 \, \text{eV}$$

We shall study forces, work, and energy in greater detail later in this chapter.

2. Position, Velocity, Acceleration, and Vectors

Position, $\vec{r}$, the location of an object is a vector, having magnitude and direction. This is illustrated in the figure.

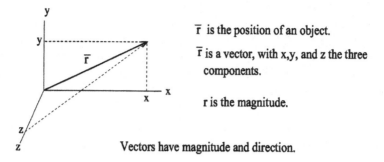

$\overline{r}$ is the position of an object.

$\overline{r}$ is a vector, with x,y, and z the three components.

r is the magnitude.

Vectors have magnitude and direction.

Velocity, $\vec{v}$ and acceleration, $\vec{a}$, are also vectors. As can be seen from the figure for position, vectors are three-dimensional, with x, y, and z components. Before we treat these important quantities as vectors we study them in one dimension.

2.1. *Position, Velocity, and Acceleration in One Dimension*

In one dimension we only must deal with the magnitude of position, r. The position of an object depends on $t = $ time. That is r is a function of t, $r(t)$. For example at $t = t_1$ the position is $r_1 = r(t_1)$. The speed, velocity in one dimension or magnitude of velocity in three dimension is defined as the ratio of the difference in positions at times t_2 and t_1, in the limit of $t_2 - t_1 \rightarrow 0$. Thus

$$v(t) = \left[\frac{r(t + \Delta t) - r(t)}{\Delta t} \right]_{\Delta t \rightarrow 0}, \tag{7}$$

or

$$v(t) = \left[\frac{\Delta r(t)}{\Delta t} \right]_{\Delta t \rightarrow 0} \equiv \frac{dr(t)}{dt} \equiv \dot{r}(t). \tag{8}$$

Note that the quantity $\frac{dr(t)}{dt}$ is called the first derivative of $r(t)$ with respect to time $= t$.

The acceleration in one dimension $a(t)$, is defined as the first derivative of $v(t)$ with respect to time $= t$:

$$a(t) = \left[\frac{\Delta v(t)}{\Delta t}\right]_{\Delta t \to 0} \equiv \frac{dv(t)}{dt} \equiv \dot{v}(t) \quad \text{or}$$

$$a(t) = \frac{d^2 r(t)}{dt^2} \equiv \ddot{r}(t). \tag{9}$$

That is acceleration, the first derivative of $v(t)$ with respect to t, is also the second derivative of $r(t)$ with respect to t.

This is illustrated in the figure below, which also illustrates a function, such as $r(t)$ as a function of t.

Position, Velocity and Acceleration are functions of time

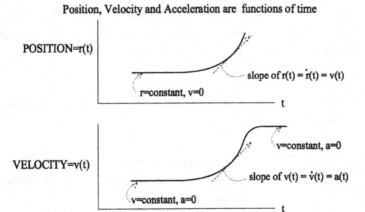

VELOCITY=TIME RATE OF CHANGE OF POSITION: v(t)= dr(t)/dt
ACCELERATION = TIME RATE OF CHANGE OF VELOCITY:
a(t) = dv(t)/dt = d²r(t)/dt²

2.1.1. *Example of Position, Velocity and Acceleration in One Dimension*

Using $d(\text{constant})/dt = 0$, $d(t)/dt = 1$, with t in units of s $=$ seconds,

$$r = \text{constant} \to v = a = 0;$$
$$r = 3\,(\text{cm/s}) \times t \to v = 3\,(\text{cm/s}); \quad a = 0; \tag{10}$$
$$v = 5\,(\text{cm/s}^2) \times t \to a = 5\,\text{cm/s}^2.$$

2.2. *Position, Velocity, and Acceleration in Three Dimensions*

Position, velocity and acceleration are vectors which can be functions of time. Thus position $= \vec{r}(t)$, velocity $= \vec{v}(t)$, acceleration $= \vec{a}(t)$, and

$$\vec{v}(t) = \left[\frac{\vec{r}(t + \Delta t) - \vec{r}(t)}{\Delta t} \right]_{\Delta t \to 0} = \frac{d\vec{r}}{dt} \equiv \dot{\vec{r}}$$

$$\vec{a}(t) = \left[\frac{\vec{v}(t + \Delta t) - \vec{v}(t)}{\Delta t} \right]_{\Delta t \to 0} = \frac{d\vec{v}}{dt} \equiv \dot{\vec{v}} = \frac{d^2\vec{r}}{d^2t} \equiv \ddot{\vec{r}}. \quad (11)$$

Note that the change in position and/or velocity with respect to time can be a change in magnitude, direction, or both. An example is a mass moving in a circle with constant speed. It experiences acceleration, called centripetal acceleration, which is very important for our study of stellar and galaxy systems.

2.2.1. *Addition of Vectors*

A vector has magnitude and direction. The addition of vectors is illustrated in the figure below.

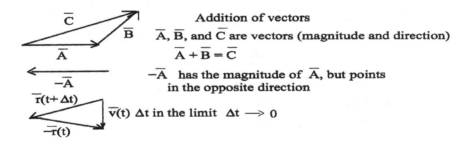

Addition of vectors

$\overline{A}$, $\overline{B}$, and $\overline{C}$ are vectors (magnitude and direction)

$\overline{A} + \overline{B} = \overline{C}$

$-\overline{A}$ has the magnitude of $\overline{A}$, but points in the opposite direction

$\overline{v}(t) \, \Delta t$ in the limit $\Delta t \longrightarrow 0$

Note that the vector C = the sum of vectors A and B does not have the sum of the magnitudes of A + B, unless A and B point in the same direction.

2.2.2. *Centripetal Acceleration*

Consider a particle moving in a circle of radius R with constant speed $= v$. Note that v is a constant, but velocity $\vec{v}(t)$ is not, and if the direction of velocity changes as time increases there is acceleration even if the speed v is constant, with a_c = centripetal acceleration

$$a_c = v^2/R, \tag{12}$$

which is derived in the figure below.

As an example, if a particle is moving with constant speed 12 cm/s along a circle of radius 2 m,

$$v = 12\,\text{cm/s}$$
$$r = 2\,\text{m} = 200\,\text{cm}$$
$$a = \frac{v^2}{r} = \frac{144\,\text{cm}^2/\text{s}^2}{200\,\text{cm}} = 0.72\,\text{cm/s}^2. \tag{13}$$

Derivation of Centripetal Acceleration

Circumference of a circle with radius R = 2πR

Proof:

D = length of arc with interior angle θ and radius R

D = R dθ, with θ in units of radians and 2π radians in a circle

Therefore, the circumference of a circle = 2πR

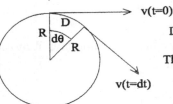

v(t=0)

D=R dθ = vdt (distance travelled in time dt with constant speed v)

Therefore d θ = $\frac{v}{R}$ dt, or $\frac{d\theta}{dt} = \frac{v}{R}$

Note magnitude of v(t=0)=v(t=dt)=v

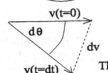

Note for dθ small dv is the arc length:

dv = v dθ

Therefore acceleration = $a_c = \frac{dv}{dt} = v\frac{d\theta}{dt} = \frac{v^2}{R}$

THIS IS CENTRIPETAL ACCELERATION

3. Force and Newton's Law of Motion

If the velocity of a mass m changes, that is if it experiences acceleration, then there is a force acting on it. Newton's famous Law of Motion:

$$\vec{F} = m \times \vec{a} \tag{14}$$

has been used for hundreds of years. As will be explained in Chapter 6, Einstein's Special and General Theories of Relativity derived a different law of motion, but for most applications Newton's law is suitable. We now consider a slightly different form which is correct for both nonrelativistic and relativistic theories.

The standard unit of force is a newton $=$ N, and of mass is a kilogram $=$ kg.

3.1. *Momentum and Newton's Law of Motion*

The nonrelativistic definition of momentum, $\vec{p}$ is [3]

$$\vec{p} = m\vec{v}. \tag{15}$$

Since m is a constant and $\frac{d\vec{v}}{dt} = \dot{\vec{v}}$, Newton's Law of Motion can be rewritten as

$$\vec{F} = \frac{d\vec{p}}{dt} = \dot{\vec{p}}. \tag{16}$$

As we shall see, Einstein's special theory has the same law of motion with a different definition of momentum.

3.1.1. *Example of a Force Constant in Magnitude and Direction*

Let us consider a force $\vec{F} = F$ in the x direction, with F a constant. Since there is only one direction, this is a one-dimensional problem and we do not need to use vectors.

Consider the case when F is acting on a particle with mass m at the position $x = x_o$, with speed $v = v_o$ at time $t = 0$. From

Newton's law:

$$F = ma = m\frac{dv}{dt} = m\frac{d^2x(t)}{dt^2} \quad \text{or}$$

$$\frac{d^2x(t)}{dt^2} = \frac{F}{m} = \text{acceleration}. \tag{17}$$

The solution to this differential equation is

$$x(t) = x_o + v_o t + (1/2)(F/m)t^2. \tag{18}$$

To prove that this is the correct solution we use

$$\frac{dt^n}{dt} = nt^{n-1}, \tag{19}$$

which includes $n = 0$ since $t^0 = 1 = \text{constant}$ and $\frac{d\,\text{constant}}{dt} = 0$. Therefore

$$x(t = 0) = x_o$$

$$v(t = 0) = \frac{dx(t)}{dt}\bigg|_{t=0} = v_o + (F/M)t|_{t=0} = v_o$$

$$a(t) = \frac{d^2x(t)}{dt^2} = 2(1/2)(F/M) = F/M = a = \text{constant},$$

$$\tag{20}$$

which shows that our solution for $x(t)$, Eq. (18) satisfies Newton's law, our initial conditions on position and speed, and the constant acceleration $a = F/M$. The usual form of the equation for position with a constant force is

$$x(t) = x_o + v_o t + \frac{1}{2}at^2. \tag{21}$$

3.2. *Newton's Force of Gravity*

The force of greatest interest to us in studying the evolution of the universe is the gravitational force. Newton's famous gravitational force for two masses m and M separated by a distance R, as illustrated in the figure, which we call F_g is

$$\text{m} \longleftarrow \text{--} \longrightarrow \text{M}$$
$$R$$

$$F_g = G\frac{mM}{R^2}, \tag{22}$$

with G being Newton's gravitational constant, $G = 6.67 \times 10^{11} \, \text{m}^3/$ $(\text{kg} \cdot \text{s}^2)$. The two masses are attracted to each other. In other words, the force on m is in the direction toward M. One example is a mass m on the surface of the earth. Using the radius of the earth R (assuming that it is a sphere, which is approximately true), the mass of the earth M, and G, the acceleration of gravity $= g = $ force of gravity on mass m divided by m is

$$g = F_g/m = G\frac{M}{R^2}$$
$$g = 9.80 \, \text{m/s}^2. \tag{23}$$

3.2.1. *Experimental Test of the Value of g*

It is very simple to test if Eq. (23) is the correct value of the acceleration of gravity at the surface of the earth. Just take an object and hold it at a height h from the ground. Drop it and measure the time t_h that it takes to fall to the ground. From Eq. (21), with $x_o = 0$, $v_o = 0, t = t_h, x(t_h) = h, a = g$, one finds

$$g = 2\frac{h}{t_h^2}. \tag{24}$$

Note that you should drop the object at least 10 times to estimate g and the error in your experiment. This is an example of testing a hypothesis by setting up an experiment, carrying out the experiment enough times to have an accurate value and an estimate of the error in your experimental arrangement. For this experiment you might consider improvements that will provide a more accurate measurement with less error.

For example, you should be careful to drop the object as close to the height h as possible, and make sure the object has no velocity when you drop it at $t = 0$.

Another interesting experiment would be to measure the speed of the object when it hits the ground, a test of conservation of energy as potential energy becomes kinetic energy, using results given later in this chapter.

3.2.2. *Earth Rotating Around the Sun*

Another example, which we will consider when treating our solar system, is the earth rotating around the sun:

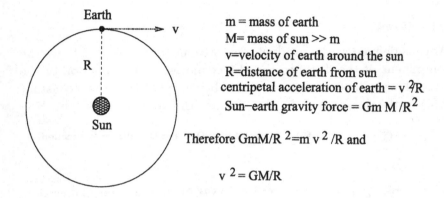

m = mass of earth
M= mass of sun >> m
v=velocity of earth around the sun
R=distance of earth from sun
centripetal acceleration of earth = v^2/R
Sun–earth gravity force = $Gm M /R^2$

Therefore $GmM/R^2 = m v^2 /R$ and

$$v^2 = GM/R$$

From this one finds that the velocity of the earth around the sun, v, is

$$v = \sqrt{\frac{GM}{R}}, \tag{25}$$

where G is Newton's gravitational constant, M = mass of sun, and R is the distance of the earth from the sun.

In a future chapter we shall see how one determines R and using the known values for G and M find that Eq. (25) is correct. That is knowing R and using v from Eq. (25) we shall show that it takes one year for the earth to move around the sun.

Another application was the discovery of Dark Matter by studying the rotation of our Milky Way galaxy and other galaxy. By using the Doppler shift, which we derive and discuss in the chapter on Hubble's Law, in the 1930s astronomers were able to measure the velocity of rotation of stars in the Milky Way. They were also able to estimate the amount of mass of visible matter. In order to provide the gravitational force to keep our galaxy rotating much more mass than the visible mass is needed. Since it is not visible it is called Dark

Matter. This is an important topic, which will be discussed in later chapters.

4. Energy and Temperature

In this section we study the various forms of energy: (1) kinetic, (2) potential, (3) heat, and work, which produces energy.

4.1. *Work*

Work is defined as the product of a force moving an object times the component of the distance the object moves in the direction of the force. This is illustrated in the figure below. That is if F is a constant force in the direction of D, the distance the object moves

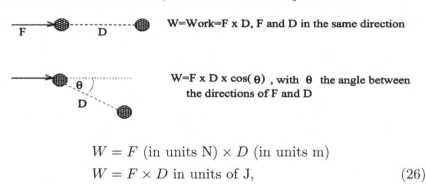

W=Work=F x D, F and D in the same direction

W=F x D x cos(θ) , with θ the angle between the directions of F and D

$$W = F \text{ (in units N)} \times D \text{ (in units m)}$$
$$W = F \times D \text{ in units of J}, \tag{26}$$

where the unit of force, the newton $= N = $ kg $\times$ m/s^2, and J $=$ N $\times$ m $=$ kg $\times$ m^2/s^2 is a joule, as defined above, the unit of energy, work, and heat.

4.2. *Potential Energy*

Potential energy is stored energy, as illustrated in the figure below.

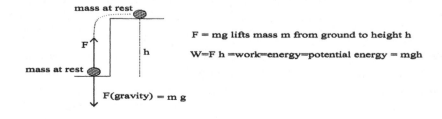

F = mg lifts mass m from ground to height h

W=F h =work=energy=potential energy = mgh

In this example the potential energy of a mass lifted a height h is mgh.

4.3. *Potential and Force*

When treating the energy of a system it is often useful to use the potential. In one dimension, say r, if a potential is $V(r)$, the corresponding force is

$$F(r) = -\frac{dV(r)}{dr}. \tag{27}$$

An example is the gravity force with masses m and M separated by a distance r, $F_g = GmM/r^2$. The gravitational potential is $V_g(r) = GmM/r$, since $-d(1/r)/dr = 1/r^2$.

4.4. *Kinetic Energy*

Kinetic energy is the energy of motion. If an object has mass m and speed v, its kinetic energy, $K.E.$, is

$$K.E. = \frac{1}{2}mv^2. \tag{28}$$

4.5. *Mass Energy*

When we study Einstein's theory of relativity we will meet another form of energy, mass energy. A mass m at rest with no potential or heat energy has the familiar mass energy

$$E = mc^2, \tag{29}$$

where $c =$ the speed of light:

$$c \simeq 300{,}000{,}000 \, \text{m/s}. \tag{30}$$

The speed of light and the distance that light moves will also be very important in our astrophysical studies.

4.6. *The Electron-Volt Energy Unit*

In our cosmological studies we shall use the electron volt, the eV, as our unit of energy: $1 \, \text{eV} =$ potential energy that an electron receives when it is raised through a one volt electric potential difference. We also use MeV (million electron volts) and GeV (billion electron volts)

$$1\,\mathrm{MeV} = 10^6\,\mathrm{eV}$$
$$1\,\mathrm{GeV} = 10^9\,\mathrm{eV}. \tag{31}$$

For example the mass energy of an electron is $m_e c^2 \simeq 0.511\,\mathrm{MeV}$, while the mass energy of a proton is $m_p c^2 \simeq 1.0\,\mathrm{GeV}$.

4.7. *Conservation of Energy*

Energy is conserved, but one form of energy can be converted to another form. For example, if there is no energy loss to friction or air resistance, if the mass m at height h falls to the ground the potential energy converts to kinetic energy:

$$\frac{1}{2}mv^2 = mgh. \tag{32}$$

Note also that potential and kinetic energy can convert to heat energy, which we now discuss.

4.8. *Temperature, Heat Energy, and Phase Transitions*

Matter has heat energy, basically the kinetic energy of motion of its constituents, such as molecules of a gas. When all of the matter is at the same temperature the system is in thermal equilibrium, and has a temperature T. The units of T are degree Celsius, °C, or kelvin K with

$$0°\mathrm{C} = 273\mathrm{K}$$
$$0\mathrm{K} = \text{absolute zero} \tag{33}$$

In the US, temperature in units of Fahrenheit, °F, are used, with

$$0°\mathrm{C} = 32°\mathrm{F}$$
$$\text{each } °\mathrm{C} = 1.8°\mathrm{F} \quad \text{or}$$
$$T_C = \frac{5}{9}(T_F - 32). \tag{34}$$

Of great importance for our study of the evolution of the universe are phase transitions, such as water, a liquid, becoming steam, a gas. This water–steam phase transition at standard atmospheric pressure occurs at the critical temperature $T_c = 100°$C. If one heats water the heat raises the temperature until it reaches the critical temperature of the water–steam phase transition. The temperature T remains at T_c until all of the water converts to steam, after which the temperature of the steam increases as the heat is added. The heat energy during the phase transition is called latent heat. Similarly, the water–ice (liquid–solid) phase transition occurs at $T = 0°$C, and the latent heat L_{w-i} (heat energy per unit of mass) is

$$L_{w-i} = 3.34 \times 10^5 \text{ J/kg}. \tag{35}$$

This will be very important for our study of the evolution of the universe, as matter as we know it was created during cosmological phase transitions. We shall need the solution to Einstein's equations in order to determine the temperature as a function of time in order to find at what time these phase transitions took place.

4.8.1. *Temperature as a Form of Energy*

Since temperature is a measure of the kinetic energy of matter in thermal equilibrium, it is a form of energy:

$$\begin{aligned} \text{Energy} &= kT \\ k &= \text{Boltzman's constant} = 8.62 \times 10^{-5} \text{ eV K}^{-1} \\ 1 \text{ eV} &= 11,604\text{K}. \end{aligned} \tag{36}$$

4.8.2. *First Law of Thermodynamics*

Thermodynamics has two important laws. We shall not need the Second Law of Thermodynamics, but the First Law is essential for our study of Einstein's equations to find $T(t)$, the temperature of the universe as a function of time from the Big Bang until now.

The internal energy of a volume of matter, called U, is the total kinetic and potential energy of the matter. If we define Q as the quantity of heat energy added to the matter, and W as the work done by the matter, the First Law of Thermodynamics states that the change in the internal energy, ΔU, is equal to the heat energy added less the work done:

$$\Delta U = Q - W. \tag{37}$$

For the study of the universe $Q = 0$, as by definition our universe is isolated, as we shall see. Also, let us define pressure, $p = $ force per area. Therefore, since work is force times distance, and if the force acts on one surface of the volume, which we take as a cube, the force is the pressure times the area, and the change in volume is the area times the distance that surface moves, so the work done on the volume is

$$W = p \times \text{change in volume}, \tag{38}$$

so the First Law of Thermodynamics for the universe is

$$\Delta U = -p \times \text{change in volume}. \tag{39}$$

4.8.3. *Equation of State*

The equation of state of a particular material relates the pressure, p on a volume of that material to various properties, such as temperature, mass, etc. For our study of the universe the equation of state has the form

$$p = w\rho, \tag{40}$$

where w is a constant and ρ is the density $=$ mass/volume. We shall use Eqs. (39), (40) and Einstein's equations to find the radius of the universe (which we define when we study Einstein's Special Theory of Relativity) as a function of time, a most important relation for understanding the evolution of the universe.

5. Dark Energy

In this chapter we reviewed many of the basic concepts that will be needed for the study of the evolution of the universe. What we will learn in Chapter 7 is that the radius of the universe, R, and the temperature, T, are functions of time

$$R = R(t)$$
$$T = T(t), \tag{41}$$

and we will derive $R(t), T(t)$ using Einstein's General Theory of Relativity.

One surprising aspect of the universe that we shall study is that most of the matter does not obey Newton's law of gravity. It is called Dark Energy. Instead of Dark Energy attracting ordinary matter, like stars, it repels. That is, Dark Energy is anti-gravity matter. It is the greatest mystery of astrophysics. We shall discuss it, and give an example of a model for Dark Energy, but there are many models and no one really knows what Dark Energy is.

As we shall discuss, Dark Matter is not understood at the present time, but is probably made of particles not in the Standard Model, a topic of the next chapter. The nature of Dark Matter will probably be discovered using modern particle accelerators within the next decade. Dark Energy, however, will probably be a mystery for many decades.

6. Problems

1. A runner, running along a straight track runs 2 miles in 8 minutes. What was his average speed in meters per second?

2a. A constant force is acting on a mass. At the initial time ($t = 0$) the mass is at rest ($v = 0$), and after 1 minute it has speed $v = 30\,\text{m/s}$. What is the acceleration?

2b. What is the speed after 2 more minutes (3 minutes after $t = 0$)?

3a. What is the circumference in meters of a circle with radius 25 meters?

3b. If a runner runs around this circumference in 1 minute, what was her average speed in meters per second?

4a. If the runner in Problem 3 runs with constant speed, what is the acceleration?

4b. At any time, what is the direction of the acceleration?

5. A vector $\vec{A}$ has components $A_x = 12\,\text{m}$, $A_y = -6\,\text{m}$. What is the magnitude and direction of $\vec{A}$? Remember if a right triangle has sides a and b, the hypotenuse $c = \sqrt{a^2 + b^2}$.

6. A vector $\vec{A}$ is 1 m long and is in the x-direction. A second vector $\vec{B}$ is also 1 m long, but is in the y-direction. What is the magnitude and direction of the vector $\vec{C}$, which is the sum of $\vec{A}$ and $\vec{B}$? That is, $\vec{C} = \vec{A} + \vec{B}$.

7a. A lead nucleus consists of 82 protons and 126 neutrons. Its mass is about 208 times the proton mass $m_p = 1.67 \times 10^{-27}$ kg. If a lead nucleus is dropped with initial speed zero from a height of 1 km, using the acceleration of gravity $= g$ for an object at the surface of earth, how much time does it take for the nucleus to hit the ground?

7b. What is the velocity of the nucleus when it hits the ground?

7c. What is the kinetic energy of the nucleus in joules when it hits the ground?

8. Noting that 1 joule of energy is related to eV by 1 J $= 6.2 \times 10^{18}$ eV, in Problem 7c what is the energy in eV? If all of this energy is turned into heat, what is the temperature T if the matter were at thermal equilibrium?

References

[1] Frank J. Blatt, *Principles of Physics*, published by Allen and Bacon, Inc.
[2] David Halliday and Robert Resnick, *Physics* Part I and Part II, published by John Wiley & Sons, Inc.
[3] Hugh D. Young, *Physics*, published by Addison-Wesley Publishing Company.

Chapter 2

Forces and Particles

We start this chapter with a review of Einstein's photoelectric effect, which used Planck's concept of quanta, particles of quantum fields, which we need to understand forces between particles.

1. Einstein's Photoelectric Effect: Quanta of Electromagnetic Field

In the year 1905 Einstein proposed the photoelectric effect. The concepts involved are essential for our discussion of forces and particles. Light travels as a wave, an electromagnetic wave. The color of light depends on the wavelength, the distance between peaks of the wave. Its speed is c, as was well-known by 1905. The color of light depends on the wavelength, λ. Red light has a longer wavelength than blue light: $\lambda(\text{red}) > \lambda(\text{blue})$.

Einstein's idea was that although light travels as a wave, it delivers energy as a particle, a quantum of the electromagnetic field, called a photon. The energy of the photon depends on the wavelength, not the brightness of the light, with $h =$ Planck's constant defined below:

$$E(\text{quantum, a photon}) = h\frac{c}{\lambda}. \tag{1}$$

This is shown in the figure. Light, a travelling wave delivers energy to a metal plate as a photon, a quantum of the electromagnetic field. If the energy of a photon is greater than the binding energy of an electron in the metal, it will cause the electron which absorbs the photon to leave the plate. Since in a typical setup many-many electrons are emited from the metal plate, and an electric current is produces by the motion of electrons, this will produce an electric current. The battery and wires needed for the current are not shown.

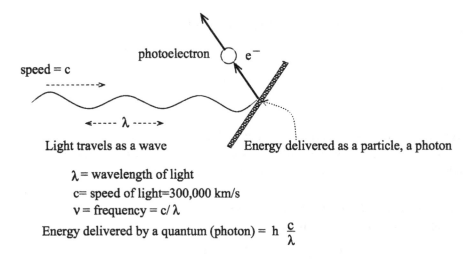

λ = wavelength of light
c = speed of light=300,000 km/s
v = frequency = c/λ
Energy delivered by a quantum (photon) = h $\frac{c}{\lambda}$

Since the energy of the photon is greater when the wavelength of the light is smaller, a photon from blue light has more energy than red light. This was Einstein's hypothesis for his photoelectric effect:

Blue light will cause a current, but red light will not, regardless of the brightness of the light.

This can be shown to be true with a rather simple experimental setup. It is essential for our understanding of forces. In classical theory the force between two particles, such as the electric force between two electrons, is caused by the electric field that each electron emits. In quantum electromagnetic field theory, a complicated theory which we need in order to understand the forces in the universe, the force is caused by an exchange of a quantum of the electromagnetic field, the photon.

This is also true for the strong field giving the strong force and the weak fields giving the weak forces. The gravity force, which is the most important for the evolution of the universe, as we shall show in Chapter 6, is more complicated, and we use classical rather than quantum theory to treat it.

Now that we have introduced the concept of the quantum of a field, let us proceed to discuss the Standard Model of elementary particles and forces, and the matter in our universe that is composed of these elementary particles.

2. Elementary Particles and Forces

Elementary particles have no components. For example, electrons are believed to be elementary. Elementary Forces are the forces between elementary particles. Particles interact via fields. See, e.g., Perkins [1] for a review of the Standard Model.

First let us consider the classical theory of forces. A particle emits a field and another particle in the field feels a force. For example, an electron has an electric charge and is surrounded by an electric field. Another electron at some distance from the first electron experiences a force, the electric force, with the strength of the force depending on the distance between the two electrons.

In quantum field theory a particle emits a quantum of an elementary field and another particle absorbs the quantum. The exchange of the quantum, which is also a particle, carries the energy of the field and produces a force on the two particles which produce and absorb the quantum of the field.

2.1. *Four Known Forces*

(1) FORCE OF GRAVITY: Our most familiar force. It drops an apple on Newton's head, and keeps the earth going around the sun. As we have learned in Chapter 1, the force of gravity between masses m_1, m_2 a distance R apart is

$$F_g = G\frac{m_1 m_2}{R^2}, \tag{2}$$

with G Newton's constant $= 6.67 \times 10^{-11}\,\mathrm{m^3/(kgs^2)}$. This is a classical theory, but we shall discuss the graviton, the quantum of the gravitational field, later.

For application to the evolution of the universe one uses Einstein's General Theory of Relativity, but this is still a classical field theory. Quantum field theory does not work for the gravitational force.

(2) ELECTROMAGNETIC FORCE: The force that holds atoms together. This is the force mainly involved in Chemistry, Biology, and Material science. Using classical field theory, the magnitude of the electric force depends on the electric charge of the interacting particles and the distance between them. For example consider two charged particles with electric charges q_1, q_2, separated by the distance R. The magnitude of the force each particle feels, called the Coulomb force, F_C, is in standard units

$$F_C = \frac{1}{4\pi\epsilon_o}\frac{q_1 q_2}{R^2},\tag{3}$$

with $\epsilon_o = 8.988 \times 10^{-12}\,\mathrm{C}^2/(\mathrm{Nm}^2)$, and C a coulomb, the unit of electric charge. As we shall show, this force is obtained from quantum field theory.

The magnetic field, $\vec{B}$, is very important for our studies of cosmology and astrophysics. The force on a particle with electric charge q moving with velocity $\vec{v}$ in a magnetic field is given by

$$\vec{F}_M = q\vec{v} \times \vec{B},\tag{4}$$

where $\times$ is the vector product equal to the magnitudes of v and B times the sine of the angle between them, and $\vec{F}$ is in the direction perpendicular to the $\vec{v}, \vec{B}$ plane. We shall not need this force in our studies.

(3) STRONG INTERACTION: This force holds quarks (elementary particles) together to make protons and neutrons (NOT elementary, as we shall discuss). One uses quantum field theory to obtain this force, but it is much more complicated than the electric force, as discussed below.

(4) WEAK INTERACTION: This force causes neutrons to decay into protons. Neutrinos, partners of electrons, as we shall see, interact only with the weak interaction. This is important for astrophysics, such as supernovae, which we shall study.

2.2. *Brief History of Particles*

Atoms were discovered in nineteenth century. In 1911 the Rutherford experiment, scattering of a beam of particles by atoms, showed most of atomic mass, concentrated in the atomic nucleus, is at the center, and at a great distance are electrons. The electric charge of the nucleus is Ze, with Z the number of electrons, each having electric charge $-e$. Z is also called the atomic number.

Atomic spectra had been observed. Using a spectrometer or even a prism it was seen that the light emitted from hydrogen was a series of lines, with colors red, blue, violet, etc. This required a new theory, as in classical physics the electrons spinning around the nucleus would emit the entire rainbow of colors. More discoveries and progress were made subsequently:

1913 Niels Bohr proposed his model of special orbits, which was the start of quantum theory.

1930–32 Atomic nuclei with atomic number Z were shown to have Z protons, each with electric charge $+e$ and neutrons, which have no electric charge.

1935–40 Yukawa postulated a theory of the nuclear force between neutrons and protons: the force comes from the exchange of mesons, like the electric force arises from the exchange of photons — discussed below.

1947 The lightest meson the pion (π) was discovered in cosmic rays.

1952 With new cyclotrons pion beams were produced. The Delta, with a mass of 1232 MeV was discovered in pion scattering on protons. The delta is an excited state of the proton. This proved that the proton is not an elementary particle, as elementary particles have no composite structure or excited states.

1953 Strange particles were discovered. The Lambda, with mass of 1116 MeV lives about 10 trillion (10^{13}) times longer than expected from the decay produced by the nuclear force. Its lifetime is about 10^{-10} s compared with the Delta's lifetime of about 10^{-23} s. There was no explanation: it was called a strange particle.

Over the next two decades the discovery of many strange, as well as "charm", "bottom", "top" particles led to the quark model, with three "generations". Quarks are elementary particles. The strong force does not change strangness while the weak force does, explaining the long lifetimes of strange particles. Similarly with charm, bottom, and top.

See Ref. [2] for an introduction to Feynman diagrams used in the next section.

3. Quantum Electrodynamics

Quantum electrodynamics (QED) is the quantum field theory of electromagnetism. The electric force is caused by exchange of photons — which are gauge bosons —, quanta of the EM field. The coupling of photons to a particle is given by electric charge of the particle. The force can be found from Feynman diagrams. The lowest order diagram for two electrons is shown in the figure below

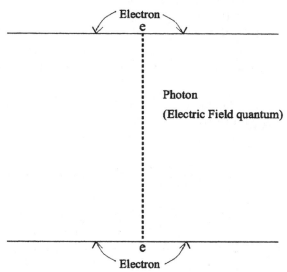

Electric Force Produced By Photon exchange

Note that $e^2 \sim 1/137$, therefore higher order diagrams are small.

ELECTRIC FORCE: From the Feynman diagram one finds that two particles with charges 1 and 2 separated by a distance $= d$ feel a force

$$\text{Force (electric)} = \frac{\text{electric charge 1} \times \text{electric charge 2}}{d \times d} \tag{5}$$

If the two charges are both positive or both negative the force repels the particles — pushes them apart. If the two charges are of opposite sign, with one positive and the other negative, the electric force attracts the two particles.

For example, consider the electric force between two electrons and between an electron and a proton, using the fact that the electric charge of an electron is $-e$, while that of a proton is $+e$. This is shown in the figure below:

$$\text{LIKE CHARGES REPEL : ELECTRIC FORCE} = F = \frac{(e)^2}{d^2} \text{ OUTWARD}$$

Electron   Electron

$$F \longleftarrow\!\!-\!\!\diamond\!\!\longleftarrow d = \text{distance} \longrightarrow\!\!\diamond\!\!\longrightarrow F$$

Charge = −e   Charge = −e

$$\text{UNLIKE CHARGES ATTRACT: ELECTRIC FORCE} = F = \frac{(e)^2}{d^2} \text{ INWARD}$$

Electron   Proton

$$\diamond\!\!\longrightarrow F \quad F\longleftarrow\!\!\diamond$$

Charge = −e   Charge = +e

$$\longleftarrow d = \text{distance} \longrightarrow$$

Thus the first order Feynman diagram reproduces the classical theory as seen in Eqs. (3) and (5). As stated above, the electric force is important in creating our universe of atoms, molecules, etc., discussed next. The coupling constant between the photon and an electron or proton, with magnitude e, or the force between two such charged particles, which is proportional to e^2, is small, as $e^2 = 1/137$. This is why the lowest order Feynman diagram, discussed above, gives the electric Coulomb force.

3.1. *Atoms: Electrons and Atomic Nucleus Bounded by the Electric Force*

Atoms are the basic components of matter in our present universe. They are formed by electrons, with charge $= -e$, bounded through the electric force to atomic nuclei, with charge Ze, where Z is the number of protons in the atomic nucleus, which we discuss below. This is illustrated in the following figure.

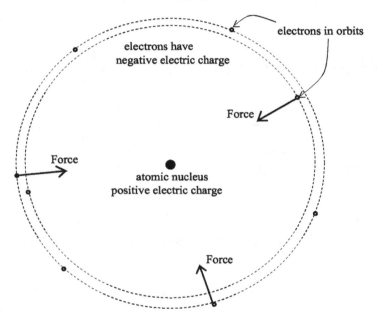

In the figure electrons are shown in orbits at a great distance from the atomic nucleus. This is the classical picture of atoms, which was developed in response to the Rutherford experiment.

In 1911 the Rutherford experiment showed that an atom consists of electrons at the known radius of an atom and most of the mass of the atom is in a much smaller component, called the atomic nucleus. Since the electrons have negative charge, and the atomic nucleus positive charge, in classical theory the electrons must be rotating about the nucleus, or the electrons would fall until they collided with the nucleus.

In the classical theory, since the electrons are moving in a circle with some radius, say r, and a velocity v the electrons have centripetal acceleration $= v^2/r$, and from the classical theory of electromagnetism they radiate electromagnetic waves (which we will study in the chapter on Hubble's discovery that the universe is expanding). From this theory one can show that the electrons would spiral down to hit the nucleus in less than a millionth of a second, and atoms would not exist. We would not exist.

It is answer to this problem that Niels Bohr developed his theory of atoms in 1913, and about ten years later quantum theory was developed to explain atoms. We briefly discuss Bohr's model of atoms, but do not attempt to discuss quantum theory.

3.2. *Bohr's Model of Atoms*

Bohr's model, formulated to be consistent with both the Rutherford experiment and the earlier observation that atoms emit electromagnetic radiation, such as light, only with certain wave lengths. In fact, we can identify which atom is emitting light by the spectrum of its radiation. Clearly these both violate classical mechanics and electrodynamics. See Ref. [3] for a detailed description of Bohr's model.

Bohr's model is based on three assumptions:

(1) Atoms have certain states, called stationary states, and electrons in those states do not radiate. They do not spiral down. They are in stationary orbits.

(2) The energy levels of these stationary states are determined by quantizing the angular momentum of the electrons in certain orbits. The angular momentum, L, of a mass m moving in a circle of radius r with speed v is $L = mvr$. Bohr's second assumption was that the stationary states are determined by

$$L = n\hbar \tag{6}$$

where $n = 1, 2, 3, \ldots$ is called the principle quantum number and $\hbar = h/(2\pi)$, called the reduced Planck's constant, is $\hbar = 1.054 \times 10^{-34}$ Js. Planck's constant h is very important in quantum theory.

(3) Bohr's third assumption is just the conservation of energy. When an electron falls from one state to another state with lower

energy, the difference in energy is emitted as a photon, the quantum of the electromagnetic field, which was introduced in 1905 by Einstein's photoelectric effect. The energy of the photon, E, is given by the frequency, $f = c/\lambda$, where c = speed of light and λ = wavelength of light, by $E = hf$. We shall discuss this in a later chapter. The third assumption is

$$hf = \Delta E, \tag{7}$$

where ΔE is the difference between the initial and final states of the electron, $\Delta E = E_i - E_f$.

Using assumptions (1) and (2), Newton's law, and the conservation of energy one finds the energy levels of the stationary states.

Consider the hydrogen atom, with one electron and one proton having charges $-e$ and e, respectively. From Newton's law, $F = ma$, using Eq. (3) with $q_1 q_2 = -e^2$, and acceleration being centripetal acceleration $a = v^2/r$ in the direction of F, one has

$$\frac{1}{4\pi\epsilon_o}\frac{e^2}{r^2} = m_e\frac{v^2}{r}. \tag{8}$$

From Eq. (8) and Bohr's first assumption, $L = mvr = n\hbar$,

$$\frac{e^2}{4\pi\epsilon_o}m_e r = (m_e v r)^2 = n^2\hbar^2 \quad \text{or}$$

$$r_n = \frac{n^2\hbar^2}{m_e e^2/4\pi\epsilon_o} = n^2 a_o, \tag{9}$$

where a_o is the radius of the lowest energy level. Using the well-known constants, $a_o = 0.53 \times 10^{-10}$ m.

From this and the classical equations for energy one can determine the energies of the atomic states. The energy, E, is the sum of the kinetic energy and potential energy, which for the hydrogen atom gives

$$E = \frac{1}{2}m_e v^2 - \frac{e^2}{4\pi\epsilon_o r}$$

$$= -\frac{e^2}{8\pi\epsilon_o r}, \tag{10}$$

where we have used Newton's law, Eq. (8). From Eq. (9) one obtains the energy levels of the states:

$$E_n = -\frac{1}{n^2}\frac{m_e}{2}\left(\frac{e^2}{4\pi\epsilon_o\hbar}\right)^2 \quad \text{or}$$

$$E_n = -\frac{1}{n^2} \times 13.6\,\text{eV}. \tag{11}$$

The figure below illustrates the energy levels and the emission of photons.

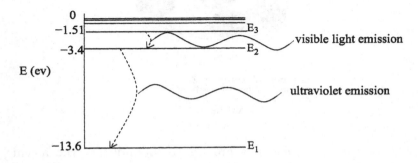

4. The Atomic Nucleus and the Nuclear Force

At the center of an atom is the atomic nucleus. It consists of Z protons and N neutrons bounded by the nuclear force. Note that the mass of an electron is $m_e = 0.511\,\text{MeV}/c^2$, while the mass of a proton is $m_p = 938.3\,\text{MeV}/c^2$ and the neutron $m_n = 939.6\,\text{MeV}/c^2$. Therefore the p and n masses are each about 2000 times that of the electron, which is why most of the mass of an atom is in the nucleus. The radius of a nucleus with N neutrons and Z protons and $A = N + Z$ is

$$R_A \simeq 1.2\,A^{1/3}\,\text{fm}, \tag{12}$$

where $1\,\text{fm} = 10^{-15}\,\text{m}$. Note that the radius of an atom is of the order of $a_o = 0.53 \times 10^{-10}\,\text{m}$, so the size of the atom is roughly 100,000 times that of its nucleus, consistent with the Rutherford experiment.

This is shown in the figure below.

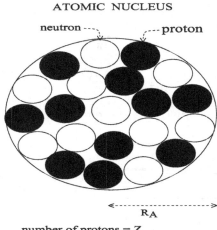

ATOMIC NUCLEUS

number of protons = Z
number of neutrons=N
A=Z+N
R_A = 1.2 x $A^{1/3}$ fm

The protons and neutrons are tightly bounded in the atomic nucleus by the nuclear force, which is much stronger than the electric force that binds electrons in atoms. The nuclear force is not an elementary force, but is very important for studies of nuclear structure. We consider it next.

4.1. *Nuclear Force*

The nuclear force is produced by the exchange of mesons, with the lighter the meson the longer the range of the force, as shown below. For example the nuclear force derived from the exchange of a pion, with the mass m_π, with the smallest mass of mesons giving the nuclear force, has a longer range than that other mesons. This is illustrated in the figure below for a neutron–proton nuclear interaction.

This is similar to the Feynman diagram for the electric force discussed above, with the exchanged photon replaced by a meson and the coupling constant g replacing e.

From this the Yukawa potential is derived, which is discussed next.

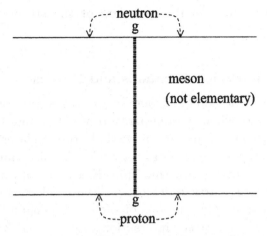

Nuclear force via meson exchange: not elementary

4.2. *Yukawa Potential*

One can find the nuclear force using the meson exchange diagram shown above. This is similar to finding the electric force with a photon exchange in QED, but neither the nucleons or meson are elementary.

The potential that one finds from the meson exchange diagram, called the Yukawa potential, is

$$V_y(r) = -g^2 \frac{e^{-\mu r}}{r}, \tag{13}$$

where g is the meson–nucleon coupling constant, similar to e for the electric force, and $\mu \propto m_m$, with m_m the mass of the exchanged meson. First note that $g^2 \simeq 100 \times e^2$, so the nuclear force is 100 times that of the electric force at short distance. Next, if m_m, the meson mass $= 0$, since $e^0 = 1$, $V_y(r) \propto 1/r$, like the electric potential discussed in Chapter 1. This is because the photon has no mass.

The lightest meson is the pion, and $\mu \propto m_\pi \simeq 1.0/\text{fm}$, which shows that the range of the nuclear force is of the order of the size of the nucleus, R_A.

The nuclear force can be found by using $F(r) = -dV(r)/dr$, but we do not need it. The elementary strong force is given by gluon exchange, where a gluon is the quantum of the strong field. We discuss this next.

5. Elementary Particles: Fermions and Bosons

We now start the discussion of elementary particles, the components of the early universe, when the temperature was too high for atoms, atomic nuclei, nucleons, or mesons, as we shall study in later chapters.

There are two types of elementary particles, fermions and bosons. Fermions have 1/2 unit of quantum spin. Electrons and quarks are fermions, as are related particles which we discuss below. Bosons have integral quantum spin. The photons of the electromagnetic interaction, the gluons of the strong interaction, and the W, Z particles of the weak interaction, called gauge bosons, have quantum spin 1, called a vector boson. The other boson in the standard model, discussed below, is the Higgs, which has 0 spin, a scalar particle.

Protons are not elementary. Neither are mesons. They are composed of quarks. As was discussed above in Section 2.2. Brief History of Particles, this was not known until the 1950s, when excited states of protons were discovered. It was not until the 1980s that there were enough experiments to derive the quark model, which we discuss in this section. The proton is composed of two u quarks and one d quark, which we discuss below, as illustrated in the figure.

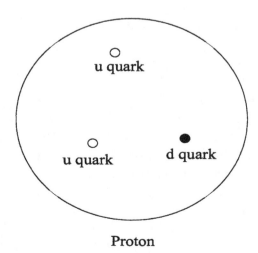

Proton

5.1. *QCD, Quantum Chromodynamics, the Strong Interaction*

Just as in QED, the strong force is given by the exchange of a quantum of the strong field, called the gluon, a gauge boson like the photon. The strong interaction between the quarks is produced by the exchange of gluons. The coupling constant between an electron and a photon, e, for electromagnetism is replaced by the quark–gluon coupling constant, g, for QCD. The lowest order Feynman diagram for the strong interaction between two quarks is illustrated below.

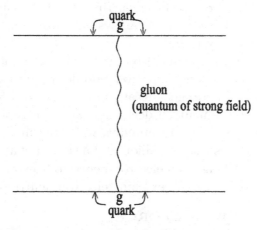

QCD (Quantum Chromodynamics):Quark force via gluon exchange

STRONG FORCE
$$g^2 \sim 1 \sim 100 \times e^2$$

Strong force ~ 100 x electromagnetic force

For QCD "color" replaces "electric charge" of QED

Since the strength of the strong coupling is about 100 times larger than the electromagnetic coupling ($g^2 \simeq 100 \times e^2$), the higher order Feynman diagrams, with more than one gluon exchanged, are as large or larger than the lowest order, depicted above. Therefore, using Feynman diagrams does not work for QCD. One must use what are called nonperturbative methods, such as large-scale computer calculations, sometimes called lattice gauge calculations.

5.2. *Weak Interaction*

Electrons do not couple to gluons, so they do not feel a strong force. They have electric charge, so they interact with photons giving an electromagnetic force. They also have a weak charge, which we call g_W, interact with the quanta of the weak field, the W^+, W^-, Z gauge bosons, and therefore feel a weak force. Neutrinos, which we discuss below, do not have either a strong or electromagnetic coupling, but do interact with the W^+, W^- and Z bosons. Since the weak interaction is much weaker than even the electromagnetic interaction, neutrinos have a very weak interaction. In fact neutrinos from the sun, which we discuss in our chapter on the solar system and galaxies, pass through us interacting very rarely.

Quarks, protons and neutrons also have weak interactions. The neutron has a larger mass than the proton, and decays into a proton, like the $n = 2$ state of hydrogen decays into the $n = 1$ state, as discussed above. Since the neutron decay is a weak process the lifetime of a free neutron is about 15 minutes, while the lifetime of an excited atomic state is less than a millionth of a second. The decay of a neutron and the scattering of an electron-neutrino from an electron are shown in the figure below. We explain "electron-neutrino" below.

WEAK FORCE

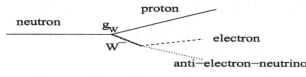

Neutron Beta Decay

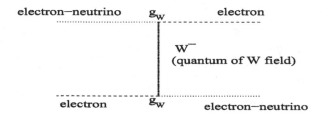

Neutrino–electron interaction via W⁻

Note that the neutron decays into a proton, an electron, and an anti-electron-neutrino. This is explained below when we discuss properties of particles and conservation principles.

6. Elementary Particles in the Standard Model

In the Standard Model of elementary particles there are fermions and bosons. We first discuss the fermions, giving their properties, and then the bosons. After this we discuss the conservation principles associated with these properties.

6.1. *Standard Model Fermions*

There are two types of Standard Model fermions, the leptons, which do not have a strong interaction, and the quarks, which do.

6.1.1. *Standard Model Leptons*

There are three generations of leptons, each with its lepton number. They are

Electron, e^- with electron number $= 1$ and $-e$ electric charge

Muon, μ^- with muon number $= 1$ and $-e$ electric charge

Tau, τ^- with tau number $= 1$ and $-e$ electric charge

The masses of the leptons are (with $m \equiv mc^2$)

$$m_e = 0.511 \, \text{MeV}$$
$$m_\mu = 105.66 \, \text{MeV} \tag{14}$$
$$m_\tau = 1,776.8 \, \text{MeV}.$$

Each lepton has a neutrino, an electron-, a muon-, and a tau-neutrino. They have no electric charge, and only have weak interactions. These neutrinos have the same lepton number as its lepton partner. E.g., the electron-neutrino has electron number $+1$.

Lepton numbers are conserved. For example, when some particle decays into an electron, like the neutron as shown above, the decay must also produce a particle with an electron number of -1, like the positron or anti-electron-neutrino. These are antiparticles, which we discuss below.

Since the muon and tau are heavier than the electron they decay into an electron, but from conservation of lepton number the decay must also include an anti-electron-neutrino. Similarly, one decay mode of the tau is

$$\tau \rightarrow \mu + \bar{\nu}_\mu + \nu_\tau,$$

with the $\bar{\nu}_\mu$ being an anti-muon-neutrino, with muon number $= -1$.

6.1.2. *Standard Model Quarks*

As discussed above, protons and mesons are composed of quarks, which are elementary particles. The quarks have electromagnetic and weak interactions, like leptons, but they also have strong interactions.

There are three generations of quarks. All quarks have baryon number 1/3. Each generation has one quark with electric charge $+(2/3)e$ and one with charge $-(1/3)e$. They are:

Up, down (u, d) with charge of $u = +(2/3)e$, $d = -(1/3)e$

Charm, strange (c, s) with charge of $c = +(2/3)e$, $s = -(1/3)e$

Top, bottom (t, b) with charge of $t = +(2/3)e$, $b = -(1/3)e$

The important property that quarks have, but leptons do not have is color. This is not color as in the rainbow, but the property of quarks that in some sense is related to the color of light, as we now explain.

6.1.3. *Strong Interaction Theory: Quantum Chromodynamics (QCD)*

Quantum chromodynamics (QCD), the theory of strong interactions, is called chromodynamics because of the color of quarks and gluons, discussed below. From QCD theory one can prove that one cannot have a free quark. An accelerator cannot make a beam of quarks like a beam of electrons or protons. That is because a quark has color and an isolated particle with color has an infinite energy.

Therefore, when we combine quarks to form a physical particle, the particle must have no color. Quarks come in three colors, which we call red (R), green (G) and blue (B). Consider a beam of colorless light. When it is bent by a glass prism the emerging light is a rainbow, which we call R plus G plus B. In the same way, if one combines $R + G + B$ light it will be colorless.

6.1.4. *Baryons*

A proton has electric charge $+e$. It is a baryon, with baryon number $= 1$. Since the u quark has charge $+2/3\,e$ and the d quark $-1/3\,e$, a proton consists of two u quarks and one d quark, with colors such that the proton has no color. Therefore

$$p = [u(R)u(G)d(B)] \tag{15}$$

$$p(\text{charge}) = \left(\frac{2}{3} + \frac{2}{3} - \frac{1}{3}\right) e = e$$

$$p(\text{color}) = 0.$$

A neutron, like a proton, is a baryon. Three quarks, making a baryon, have baryon number 1, as each has baryon number $1/3$. The neutron is composed of two d and one u quarks, with zero electric charge.

6.1.5. *Mesons*

As we discuss next, each fermion has an antiparticle. An electron has a positron. A neutrino has an antineutrino. A quark has an antiquark. The masses of antiparticles are the same as the particles, but other properties such as electric charge, lepton number, baryon number, and color are the opposite for the antiparticle. For example the positron, e^+, has charge $= +e$ and electron number -1, so if an electron and a positron collide they produce particles with no electron number or charge, such as $e^- + e^+ \rightarrow$ two photons, with charge $= 0$.

Mesons consist of a quark and an antiquark, For example the π^0

$$\pi^0 = q(R)\bar{q}(R) + q(G)\bar{q}(G) + q(B)\bar{q}(B), \tag{16}$$

where $\bar{q}$ is an antiquark with baryon number $-1/3$, as the baryon number of a meson is 0, and by q the sum over both an u and a d

quark is implied. As we have discussed there are three generations of quarks. The charm quark, c, and bottom quark, b, have masses (see next subsection) about 1.5 and 5 times the proton mass, and are often called heavy quarks. The mesons made of c, b quarks and antiquarks are very important for studies of the nature of QCD.

6.1.6. *Quark Masses*

We cannot give the masses of the quarks, as they do not exist as free particles. But the rough estimate is that the (u, d) have a few MeV, the s about 150 MeV, the c about 1,500 MeV, the b about 5,000 MeV. Since particles made of top quarks have not yet been discovered, it is difficult to estimate the mass of the t quark, but it has a mass of more than 100 protons.

7. Summary

7.1. *Antiparticles*

Every Fermion, lepton and quark, has an antiparticle.

Antiparticles have opposite charge, color, lepton number, baryon number of particle.

Lepton example: electron, electron number = 1, charge = $-e$
positron (antielectron), electron
number = -1, charge = e
Baryon number: a quark has baryon number = 1/3
an antiquark has baryon number = $-1/3$
Color: quarks have three colors: R(red), G(green), B(blue)
antiquarks have anticolors: $\bar{\text{R}}$, $\bar{\text{G}}$, $\bar{\text{B}}$
$q\bar{q}$ (meson) has color = 0

Electric charge, color, lepton number, baryon number are conserved.

Particles and antiparticles can be produced together. E.g.,

a photon can become an electron–positron

a gluon can become a quark–antiquark

a Z^0, weak field quantum, can become a neutrino–antineutrino

photon   e   — e⁻, electron
— e⁺, positron (antielectron)

gluon   g   — q, quark
— q̄, antiquark

Z^0 g_W — ν, neutrino
— ν̄, antineutrino

The interaction strengths are given by g–strong, e–electric, and g_W–weak.

Note that each lepton (electron, muon, tau) has its neutrino, with lepton number = +1 (electron, muon, tau number).

7.2. *Elementray Particles and Forces in the Standard Model*

1. STANDARD MODEL FERMIONS (quantum spin = 1/2)

	leptons	quarks
First Generation	$\begin{pmatrix} e^- \\ \nu^e \end{pmatrix}$	$\begin{pmatrix} u \\ d \end{pmatrix}$
Second Generation	$\begin{pmatrix} \mu^- \\ \nu^\mu \end{pmatrix}$	$\begin{pmatrix} c \\ s \end{pmatrix}$
Third Generation	$\begin{pmatrix} \tau^- \\ \nu^\tau \end{pmatrix}$	$\begin{pmatrix} t \\ b \end{pmatrix}$

2. STANDARD MODEL GAUGE BOSONS (quantum spin = 1)
GLUON — QUANTUM OF STRONG FIELD (QCD)
PHOTON — QUANTUM OF ELECTROMAGNETIC FIELD (QED)
W^+, W^-, Z^0 — QUANTA OF WEAK FIELDS

3. HIGGS — QUANTUM OF HIGGS FIELD — Scalar boson (quantum spin $= 0$)
   Higgs is important for the electroweak phase transition, studied below.
4. GRAVITY (GRAVITON — quantum spin $= 2$)
   Not Part of Standard Model, as quantum field theory does not work. Gravity is the most important force for the evolution of the universe. One uses Einstein's Special and General Theories of Relativity, which we study below.

Note that the three generations of leptons are called FLAVOR leptons: electron, muon, tau flavors. A neutrino with good flavor does NOT have a definite mass. This causes a neutrino of one flavor at time t to convert partly to the other flavors at later time: NEUTRINO OSCILLATION.

Neutrino oscillations are important for studying symmetries, as will be discussed below.

Beyond the Standard Model:

SUPERSYMMETRY (SUSY): Every particle has a supersymmetric partner. Recall that fermions have quantum spin $1/2$ and bosons have integral spin. The SUSY partner of a boson has half-integral spin, and that of a Fermion has integral spin. It is likely that a SUSY particle is Dark Matter (see below).

For the Electroweak Phase Transition (EWPT) the existence of SUSY particles is essential for the EWPT to be first order and mass generation and baryogenesis to work, as we shall see in Chapter 9.

STERILE NEUTRINOS: In addition to the three generations of neutrinos discussed above, sterile neutrinos have been recently discovered. They have no interaction except gravity, not even the weak interaction of standard neutrinos. It is possible that sterile neutrinos could be Dark Matter, discussed below and in our chapter on galaxies; and might cause pulsar kicks, discussed in Chapter 5.

DARK MATTER: As we shall see, from properties of galaxies and from the Cosmic Microwave Background Radiation (CMBR) about 23% of matter in the universe is Dark Matter. It cannot be composed of Standard Model particles, all of which have some interaction in addition to gravity, while Dark Matter particles have only the gravitational interaction.

DARK ENERGY: About 73% of matter in the universe is Dark Energy, which is vacuum energy. Its only force is anti-gravity. It repels standard and dark matter rather than attracting it. We shall discuss Dark Energy when treating the CMBR and supernovae, and examine a possible model.

8. Problems

1. Yellow light has a wavelength $\lambda = 5.89 \times 10^{-7}$ m. Using Planck's constant $h = 6.63 \times 10^{-34}$ Js and the speed of light $c = 3 \times 10^8$ m/s, how much energy in units of eV does a photon of yellow light deliver to a metal plate in the photoelectric effect?
2. What is the wavelength of light emitted as an electron in the $n = 2$ level of a hydrogen atom falls to the $n = 1$ level?
3. The mass of the sun $= 1.99 \times 10^{30}$ kg and the mass of the earth is 5.97×10^{24} kg. Using the average distance from the earth to the sun $= 1.5 \times 10^{11}$ m, what is the force of gravity between the earth and the sun?
4. What is the acceleration of gravity 9000 km above the surface of the earth?
5a. An electron-neutrino plus a neutron become a proton and an electron. The mass of a neutron is $m_n c^2 = 939.565$ MeV and a proton is $m_p c^2 = 939.272$ MeV. Assuming that the initial neutrino and n and the final p have zero kinetic energy, what is the energy of the electron?
5b. Assuming that it is not relativistic, what is the velocity of the electron?
6a. Draw the Feynman diagram for an electron interacting (weakly) with a proton to create an electron neutrino and a neutron. What gauge boson was exchanged?
6b. Draw the Feynman diagram for a neutrino scattering from a proton. What weak gauge boson was exchanged?

References

[1] Donald H. Perkins, *Introduction to High Energy Physics*, published by Addison-Wesley Publishing Company, Inc.
[2] Claude Itzykson and Jean-Bernard Zuber, *Quantum Field Theory*, published by McGraw-Hill, Inc.
[3] Robert Eisberg and Robert Resnick, *Quantum Physics*, published by John Wiley & Sons.

Chapter 3

Hubble's Law: Expansion of the Universe

1. Overview of Hubble's Law

A galaxy consists of a large number of stars held together by gravity. Our sun is in the Milky Way galaxy, which we discuss in the next chapter. With many stars, galaxies are very bright and can be seen at great distances from us.

In the 1920s Hubble and others carried out observations on distant galaxies from which was made one of the greatest discoveries in astrophysics: our universe is expanding. In the present chapter we review the concepts needed to understand Hubble's discovery, show his results and more recent results which lead to Hubble's Law, and discuss the implication for our study of astrophysics and the evolution of the universe. First we give the history of this important discovery.

1.1. *History of Hubble's Law*

In 1912 Vesto M. Slipher, working under the direction of Percival Lowell, carried out the spectroscopy of radiation from distant galaxies. He found that the atomic lines were not at the same wavelengths as the lines from the same atoms when measured on earth. They were REDSHIFTED. Recall that in our previous chapter we saw that each atom emitted light with a certain spectrum. From this spectrum we can tell which atom is radiating. We learn in this chapter how one can use the modification of the spectrum to measure the speed of an atom.

In the 1920s Edwin Hubble with the help of others gathered information from many galaxies. From the amount of redshift he could find the velocity of each galaxy with respect to the earth. By using information on the distance of each galaxy from us, he was able to plot the velocity as a function of distance. The velocity only depends on distance, not direction.

The result is HUBBLE'S LAW, one of the greatest discoveries in astrophysics, which we now review.

1.2. *Waves*

We are all familiar with waves. You are familiar with water waves, such as seen on an ocean beach or on the surface of a lake. Probably the first waves you experienced were sound waves, although you did not know that sound is a wave. These are travelling waves. Sound is a longitudinal wave, with the medium, such as air, being compressed and expanded in the direction of the motion of the wave. Water waves are transverse waves, with the motion of the water upward and downward, perpendicular to the direction of the wave motion.

Electromagnetic waves, are very important. The atomic radiation from galaxies, which Hubble studied, is an electromagnetic wave, a transverse wave. It is called an electromagnetic wave because the wave consists of electric and magnetic fields ($\vec{E}$ and $\vec{B}$) perpendicular to each other, and a transverse wave as both the $\vec{E}$ and $\vec{B}$ fields are perpendicular to the motion of the wave. This is illustrated in the figure on the next page.

The wavelength λ and frequency $\nu = c/\lambda$ of light change if you are moving toward or away from the source of light. This is the Doppler shift, which Hubble used to determine the speed of distant galaxies. We now study the Doppler shift, for which a little of Einstein's Special Theory of Relativity is needed.

1.3. *Doppler Shift*

The derivation of the Doppler shift, the change in the wavelength and frequency of light if the source is moving toward or away from you, needs Einstein's Special Theory of Relativity.

Light is an electromagnetic wave

speed of light = c

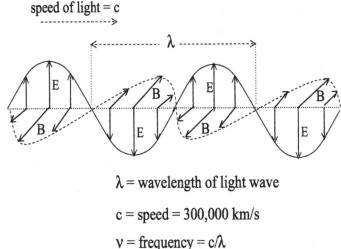

λ = wavelength of light wave

c = speed = 300,000 km/s

ν = frequency = c/λ

E = electric field

B = magnetic field

1.3.1. *Einstein's Special Theory of Relativity and Time Dilation*

The main difference between the Special Theory of Relativity and Newton's Laws of Motion is due to Einstein's axiom that the speed of light is the same in any non-accelerating system [1]. We will study Einstein's theories of relativity in detail in a future chapter, but need the effect of motion on time intervals for the derivation of the Doppler shift. We do not prove the relation here.

If a clock at rest in a frame measures a time interval Δt_o and you measure the time interval Δt of the clock which is moving with speed u with respect to you, your measurement Δt is related to the time interval Δt_o by

$$\Delta t = \Delta t_o \frac{1}{\sqrt{1 - \frac{u^2}{c^2}}}. \tag{1}$$

Therefore a moving clock has a larger time interval. For instance, if an astronaut travels in a rocket moving near the speed of light for ten years, then quickly turns back toward earth and takes ten years to return to earth, her twin brother who has aged 20 years will find his twin sister (the astronaut) much younger than he is. This time interval change is a real physical effect. For example, if a π^+ meson produced in an accelerator is moving near the speed of light its lifetime is much longer and it goes much farther before it decays than predicted by the lifetime at rest. This is important for designing particle detectors.

1.3.2. *Derivation of the Doppler Shift*

Consider an electromagnetic wave with a frequency ν_o and wavelength λ_o, with $\nu_o = c/\lambda_o$, as we have seen. The period of the wave, τ_o, is the inverse of the frequency, or $\tau_o = 1/\nu_o$. If the source of the radiation is moving toward an observer with speed v_s, we use Eq. (1) from relativity theory for time intervals to obtain the period τ' of the wave as measured by the observer at rest

$$\tau' = \frac{\tau_o}{\sqrt{1 - v_s^2/c^2}}. \tag{2}$$

Also the distance that the source moves in the time interval τ_o is $L = v_s\tau_o \rightarrow L' = v_s\tau' = (v_s\tau_o)/\sqrt{1 - v_s^2/c^2}$ for the observer from whom the source is moving with speed v_s. This reduces the period measured by the observer toward whom the source is moving with speed v_s by the time L'/c. Therefore the period of the wave from the source moving toward the observer with speed v_s is

$$\tau = \tau' - \frac{L'}{c} = \frac{\tau_o}{\sqrt{1 - v_s^2/c^2}}(1 - v_s/c)$$

$$= \tau_o\sqrt{\frac{1 - v_s/c}{1 + v_s/c}}. \tag{3}$$

Using $\lambda = c/\nu = c\tau$ this gives a blueshift (shorter wavelength)

$$\lambda = \lambda_o\sqrt{\frac{1 - v_s/c}{1 + v_s/c}}, \tag{4}$$

for the source moving toward the observer, and if the source is moving away from the observer a redshift (longer wavelength)

$$\lambda = \lambda_o \sqrt{\frac{1 + v_s/c}{1 - v_s/c}}, \tag{5}$$

where we use $v_s \rightarrow -v_s$. This is the Doppler shift, a redshift if the galaxies are moving away from us, shown in the figure below.

DOPPLER SHIFT

System of light with source at rest

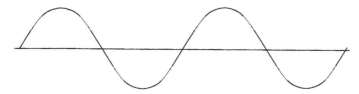

λ_0 = wavelength of light wave ν_0 = frequency = c/λ_0

Move toward the light source with speed u $\xrightarrow{\quad u \quad}$

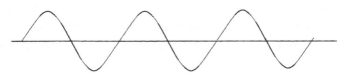

λ= wavelength of light wave $= \sqrt{\dfrac{1-(u/c)}{1+(u/c)}}\lambda_0$

(shorter wavelength)

Move away from light source with speed u $\xleftarrow{\quad u \quad}$

λ= wavelength of light wave $= \sqrt{\dfrac{1+(u/c)}{1-(u/c)}}\lambda_0$

(longer wavelength)

REDSHIFT

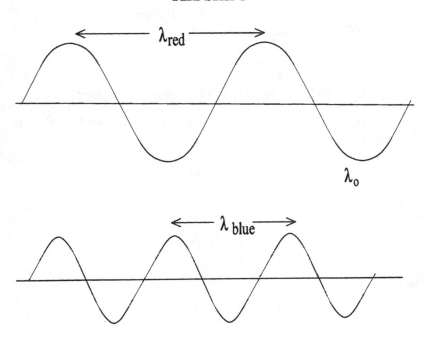

λ_{red} is larger than λ_{blue}

Therefore if a star is moving away from you light from hydrogen or any other atom in the star is redshifted.

1.3.3. *How Hubble Measured Galaxy–Earth Velocity*

We can measure the velocity of a distant galaxy relative to us using th Doppler shift by studying the radiation spectrum of hydrogen.

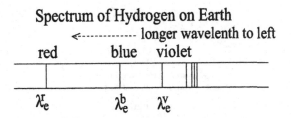

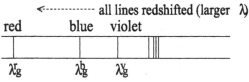

λ_e = wavelength measured with Hydrogen on Earth

λ_g = wavelength measured with Hydrogen on Galaxy

Assuming the galaxy is moving from earth with speed v_g, then by Doppler shift,

$$\lambda_g = \sqrt{\frac{1 + (v_g/c)}{1 - (v_g/c)}}\lambda_e$$

Therefore

$$\lambda_g^2 = \frac{1 + (v_g/c)}{1 - (v_g/c)}\lambda_e^2$$

Solve for v_g:

$$v_g = \frac{\lambda_g^2 - \lambda_e^2}{\lambda_g^2 + \lambda_e^2}c$$

That is how Hubble measured v_g using Doppler shift.

2. Hubble's Graphs and Hubble's Law

Hubble's measurements of the redshifts of galaxies was published in 1929 [2] as shown in the figure.

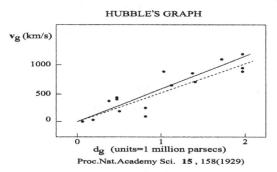

HUBBLE'S GRAPH

v_g (km/s)

d_g (units=1 million parsecs)

Proc.Nat.Academy Sci. **15** , 158(1929)

parsec = astronomical unit of distance
1 parsec = 1pc = 3 X 10^{18} cm = 3.25 light years
1 light year = distance light travels in one year
speed of light =c=300,000 km/sec

Hubble used doppler shift of known spectral lines to find v_g

Much data was collected since then, as shown below.

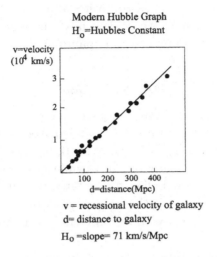

Modern Hubble Graph
H_o=Hubbles Constant

v = recessional velocity of galaxy
d= distance to galaxy
H_0 =slope= 71 km/s/Mpc

From Hubble's graph in 1929 and the more recent one shown above one finds that the velocity of a galaxy v_g is linearly related to the distance of the galaxy from earth d_{eg}:

$$v_g \propto d_{eg}. \tag{6}$$

This leads to Hubble's Law:

HUBBLE'S LAW

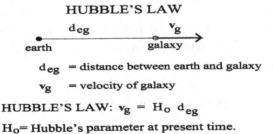

d_{eg} = distance between earth and galaxy

v_g = velocity of galaxy

HUBBLE'S LAW: $v_g = H_o\, d_{eg}$

H_o= Hubble's parameter at present time.

Hubble's law means that a galaxy at twice the distance will be moving away at twice the speed.

In general, Hubbles's perameter $= H(t)$ is a function of time and the time dependence of $H(t)$ is very important, as we shall see.

Hubble's discovery that galaxies were moving away from us is one of the most important one in astrophysics and cosmology. After many years of discussion, which we review next, it leads to our present understanding that the universe is expanding.

2.1. *Interpretations of Hubble's Law*

In the 1920s and 1930s most astronomers and other scientists did not accept the idea that the universe expanded from a very early time, for that makes one consider the time $t = 0$, and ask what happened before $t = 0$. Until the study of the Cosmic Microwave Background Radiation (CMBR) many believed in a "steady-state" universe, with matter being created to fill the void caused by the expansion. We will not discuss the "steady-state" universe, as it no longer has any possible validity.

The present interpetation is that the universe is expanding. After the evidence presented by the CMBR, there is no alternative.

From the latest CMBR observations, which we shall discuss in detail in Chapter 8, one finds that the Big Bang occured about 14 billion years ago. I.e., OUR UNIVERSE IS ABOUT 14 BILLION YEARS OLD.

We have a great deal of astrophysics and cosmology to cover before discussing the CMBR and the many properties of the universe, in addition to its age, that we can learn from studying the light from the early universe.

Many still ask: what happened before $t = 0$? They also ask: are there other universes in addition to ours? As we shall see, there are good reasons to ask these questions.

Next we discuss how Hubble's Law can be used to learn some of the most important properties of the universe.

2.2. *Hubble's Law, Hubble's Parameter, and the Universe*

Hubble's Law holds for any time, and defines $R(t)$, the radius of the universe at any time t. See, e.g., Ref. [3]. For a galaxy or any matter at a distance $R(t)$, the earth–galaxy distance in Hubble's Law, d_{eg}, becomes

$$d_{eg} \rightarrow R(t), \tag{7}$$

and the speed of the galaxy, v_g becomes the speed of expansion of $R(t)$:

$$v_g \rightarrow \dot{R}(t) = \frac{dR(t)}{dt}. \tag{8}$$

From this Hubble's parameter, and Hubble's Law become

$$H(t) = \frac{\dot{R}(t)}{R(t)}, \quad \text{and} \quad \dot{R}(t) = H(t)R(t). \tag{9}$$

Note that $\dot{R}(t)$ is the speed of expansion of the radius of the universe. From Hubble's Law, Eq. (9), with $H(t)$ fixed, we know as $R(t)$ gets larger $\dot{R}(t)$ increases. As discussed above in our brief treatment of Einstein's Special Theory of Relativity, the velocity of light $= c$ is the largest velocity possible in our universe. Anything with speed greater than c must be outside our universe in the sense that we cannot detect it. Thus the radius of our universe is given by the condition

$$\dot{R}(t) = H(t)R(t) = c \quad \text{for } R(t) = \text{radius of universe}$$
$$R(t) = \frac{c}{H(t)} = \text{radius of universe, time} = t. \tag{10}$$

Therefore, the radius of the universe at any time, called the Hubble radius, is given by the inverse of the Hubble parameter, $1/H(t)$. We shall need these concepts when we study the very early universe and inflation.

2.3. *Redshift Parameter*

The redshift parameter, z, is used by astrophysicists as a convenient measure of the distance of an astrophysical object, such as a star, galaxy, or supernova that is emitting light. It is also related to the time and therefore temperature of a cosmological event.

The definition of z is that if an object that is receding from us is emitting electromagnetic radiation, such as light, with a wavelength

λ_o that is measured by us to have wavelength λ then

$$z = \frac{\lambda - \lambda_o}{\lambda_o}. \tag{11}$$

The relationship between the redshift parameter, the recessional velocity of the object emitting electromagnetic radiation, and its distance is given in the table below. It is assumed that the Hubble constant is $H_o = 71\,\text{km/s/Mpc}$, as stated earlier.

Redshift z	Recessional velocity v/c	Distance Mpc	Distance 10^9ly
0	0	0	0
0.1	0.095	394	1.29
0.5	0.385	1,540	5.02
1.0	0.600	2,370	7.73
2.0	0.800	3,170	10.3
3.0	0.882	3,520	11.5
4.0	0.923	3,710	12.1
5.0	0.946	3,830	12.5
10.0	0.984	4,040	13.2
Infinite	1.0	4,190	13.7

Note that $z = $ infinity corresponds to $t = 0$, and since the universe is 13.7 billion years old (as we shall see), the distance is 13.7 ly.

THE EVOLUTION OF THE UNIVERSE (OVERVIEW)

$t = $ Time	$T = $ Temperature	Events
10^{-35} s	10^{14} GeV	Big Bang, INFLATION: Very early, current particle theory no good
10^{-11} s	100 GeV	Electroweak phase transition: Particles get masses (Higgs mechanism). Particle theory ok. Baryogenesis? (more particles than antiparticles)
10^{-5} s	100 MeV	QCD (quark-hadron) phase transition: Quarks (elementary particles) condense to protons
$1-100$ s	1.0×10^9 K	Nucleosynthesis: Helium, light nuclei formed, superconducting Universe

(Continued)

(*Continued*)

t = Time	T = Temperature	Events
380,000 years	0.25 eV, 3,000K	Atoms (electrically) neutral Last scattering of light (electromagnetic radiation) from Big Bang: Cosmic Microwave Background
1 billion years		Early galaxies form
14 billion years	2.7K	Now

3. Problems

1a. A π^+ meson has a lifetime of 2.6×10^{-8} s when it is at rest. An accelerator creates a π^+ meson with a velocity $v = 0.9\,c$. What is its lifetime in the system of the accelerator?

1b. How far does the pion travel before it decays?

2. A line in the spectrum of a hydrogen atom has a wavelength $\lambda = 1.3 \times 10^{-7}$ m. If the source of the radiation is moving away from the observer with a speed $v = c/2$, what wavelength is observed?

3. If a galaxy with a redshift $z = 5$ is emitting the radiation in Problem 2. ($\lambda = 1.3 \times 10^{-7}$ m if the atom is at rest), what wavelength is observed? See table of v/c vs z.

4. A rocketship approaching the earth emits a sodium line, with wavelength $\lambda = 5.896 \times 10^{-7}$ m. Find the wavelength measured on earth if the rocketship's speed is $0.1\,c$, $0.4\,c$, and $0.8\,c$.

References

[1] Robert Resnick, *Introduction to Special Relativity*, published by John Wiley & Sons, Inc.

[2] Edwin Hubble, *Proceedings of the National Academy of Sciences* **15**(3), 168–173 (1929).

[3] Roger A. Freedman and Willian J. Kaufmann III, *Universe*, published by W.H. Freeman and Company.

Chapter 4

Stars, Galaxies, Etc.

1. Overview of Chapter 4

In Chapter 4 we discuss the universe starting from a time about one billion years after the Big Bang, when stars and galaxies started to form, to the present time.

(1) Review of the force of gravity and the concept of centripetal acceleration.
(2) Find the distance from earth to the sun $= 1$ astronomical unit (A.U.).
(3) Prove using Newton's Law of Motion that the time for earth to go around the sun is one year.
(4) Nebular Theory started by René Descartes.
(5) The evolution of our solar system. Why is it a flat disk?
(6) Asteroids and comets.
(7) Our sun is a nuclear fusion plant.
(8) Galaxies and Dark Matter.
(9) Gravitational collapse of a massive star and creation of pulsars, black holes, and supernovae.

2. Review of Force of Gravity

As we saw in Chapter 1, $F_g =$ gravity force on mass m at a distance R from mass M, as shown in the figure, is $(G = 6.67 \times 10^{-11} \, \text{m}^3/(\text{kgs}^2))$

$$F_g = G \frac{mM}{R^2}. \tag{1}$$

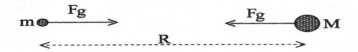

The force of gravity plays the major role in forming astronomical structures in the universe.

Starting about 1 billion years after the Big Bang gravity collapsed primary and secondary cosmic dust to form galaxies. Within the galaxies being formed gravity collapses rotating clumps of cosmic dust to form stars with planets rotating about the star, and moons rotating about the planets, like our solar system.

Stars like our sun are nuclear furnaces, with gravity pulling the fusing atomic nuclei together. Our sun has been burning for about 5 billion years, and will burn out in about 5 billion more years. Stars more massive than our sun quickly burn up their nuclear fuel and undergo gravitational colapse. This process creates supernovae. Supernovae play an important role in the universe, creating heavy atomic nuclei (secondary cosmic dust), pulsars and black holes.

An important quantity is the distance of the earth from the sun. It is 1 A.U., which we used previously. We now give the derivation.

TO MEASURE THE DISTANCE FROM THE EARTH TO THE SUN

Traditional method. Pick a star with distance D from earth known

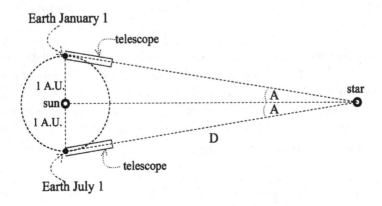

Measure the angle $2A$ using a telescope aimed at a distant star. $D =$ distance from earth to star is known.

1 A.U./$D = \sin(A) \sim A$ (very small).

1 A.U. = distance from earth to sun $\sim D \times A = 1.5 \times 10^{11}$ m

Modern method. Use radar to measure the distance of earth from the planet Venus. Use trignometry and measured angles of Earth and Venus with respect to the sun and one can accurately determine the distance of the earth from the sun, with average value = 1 A.U.

Force of gravity attracts earth to the sun.

QUESTION: Why does it not fall?

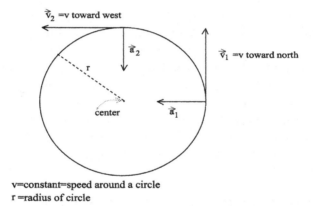

v=constant=speed around a circle
r =radius of circle

$$a = \text{centripetal acceleration} = v^2/R$$

ANSWER: Earth's gravitation acceleration = its centripetal acceleration

From this we show that it takes one year for the earth to circle the sun. Use Newton's second law: $F =$ mass $\times$ acceleration and the fact that an object moving in a circle with constant v feels a centripetal acceleration $= v^2/R$. Newton's Law is that force equals mass times acceleration. The force is the force of gravity, $F_g = G\frac{mM}{R^2}$, and acceleration is centripetal acceleration, $a_c = \frac{v^2}{R}$, therefore

$$G\frac{mM}{R^2} = m\frac{v^2}{R} \quad \text{or} \quad v^2 = MG/R. \tag{2}$$

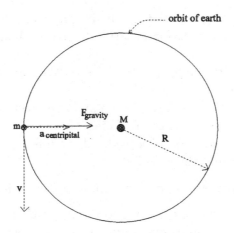

The parameters needed to find v are $G = 6.67 \times 10^{-11} \, \text{m}^3/(\text{kgs}^2)$, as we have seen, $M = $ mass of the sun $= 1.99 \times 10^3 \, \text{kg}$, and $R = $ radius of the earth's orbit $= 1 \, \text{A.U.} = 1.5 \times 10^{11} \, \text{m}$, as we have shown. Therefore,

$$v = 3 \times 10^4 \, \text{m/s}. \tag{3}$$

The distance that the earth travels around the sun each year is the circumference of the circle of radius R:

$$\text{distance} = 2\pi R. \tag{4}$$

Therefore the time for the earth to go around the sun, $t_{es} = $ the distance divided by the speed is

$$t_{es} = \frac{2\pi R}{v} = 3.15 \times 10^7 \, s = 1 \text{ year.} \tag{5}$$

3. Nebular Theory of Our Solar System — René Descartes, 17th Century

MODEL OF SOLAR SYSTEM

Start with a cloud of interstellar gas.

Gravity collapses cloud to form the sun at center.

In outer regions planets, moons, etc. are formed.

MODEL'S OBSERVATIONAL REQUIREMENTS:

Planets isolated, not bunched together.

Planet orbits are nearly circular and in one plane.

All planets rotate about the sun in the same direction as the sun's rotation, and the axis of rotation for most planets is the same as the sun's.

Most axes of rotation for most moons are the same as their parent planets'.

Also many asteroids (mini-planets) have been observed with orbits in the same plane as the planets.

Comets have also been observed. They differ from asteroids in their composition and their orbits.

QUESTION: Why are our solar system and galaxy flat disks?

ANSWER: This is due to gravity and the fact that rotating primary and secondary matter and centripetal acceleration are perpendicular to the axis of rotation.

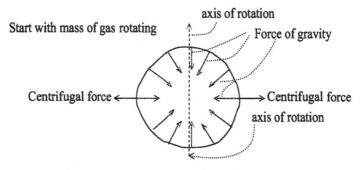

Centrifugal force balances gravitational force when perpendicular to rotation

No centrifigal force along axis of rotation.

Gravitational force collapses gas mass along rotational axis to form a disk.

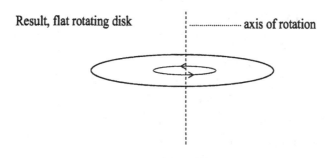

3.1. *Early Stages of Evolution of Our Solar System*

1. Solar nebula before
   condensation. Primary
   and secondary gas,
   dust.

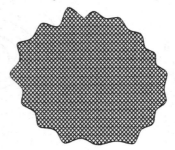

gas–dust cloud, disk rotates

2. Solar nebula contracts
   and flattens into a
   spinning disk.

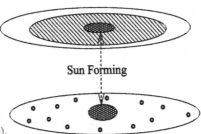

Sun Forming

3. Dust grains condense
   to form moon-size
   preplanets (planetesimals).

3.2. *Later Stages of Evolution of Our Solar System*

4. Strong winds from
   evolving sun expel
   nebular gas.

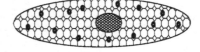

5. Planetesimals collide
   and grow.

6. After about 100×10^6 years
planetesimals form a
few planets which move
in approximately circular
orbits.

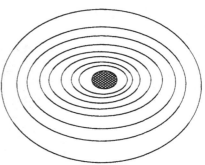

3.2.1. *Asteroids*

In addition to the planets in our solar system there are many very
small planet-like bodies, called asteroids. At the present time more
than 50,000 have been discovered, and many more are seen every
year.

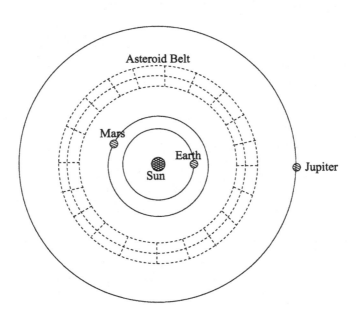

3.2.2. *Comets*

Also in our solar system there are comets. They are small collections of ice and rocks, with highly elliptical orbits, so some comets' orbits come close to the sun and then very far, near the edge of our solar system, as shown in the figure.

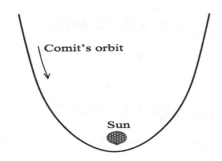

3.3. *Evolution of Our Sun Over 10 Billion Years*

See, e.g., Chaisson and McMillan [1] for definitions and discussions of our solar system, galaxies, quasars, neutron stars and black holes.

Steady State: Inward pull of gravity balanced by outward pressure of hot solar matter.

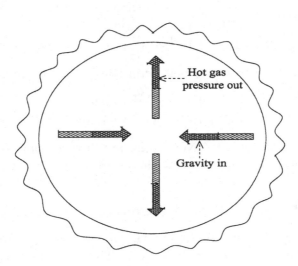

3.3.1. *Sun's Composition Over 10 Billion Years*

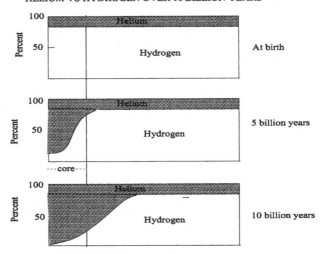

HELIUM VS HYDROGEN OVER 10 BILLION YEARS

3.3.2. *10 Billion Years of Nuclear Energy*

Outward pressure from nuclear reactions is balanced by inward pressure of gravitational attraction.

Nuclear energy starts with proton-proton reactions, with H^1 = proton, H^2 = deuterium, He^3, He^4 = Helium-3 and normal helium:

$$H^1 + H^1 \rightarrow H^2 + \text{positron} + \text{neutrino}$$
$$H^1 + H^2 \rightarrow He^3 + \text{energy}$$
$$He^3 + He^3 \rightarrow He^4 + H^1 + H^1 + \text{energy},$$

followed by helium fusion to beryllium (Be^8) and carbon (C^{12})

$$He^4 + He^4 \rightarrow Be^8 + \text{energy}$$
$$Be^8 + He^4 \rightarrow C^{12} + \text{energy}.$$

Another chain is the CNO Cycle, in which carbon serves as a catalyst. Without going through the entire chain, it can be represented by:

$$C^{12} + 4(H^1) \rightarrow C^{12} + He^4 + 2 \text{ positrons} + 2 \text{ neutrinos} + \text{energy}.$$

Finally, after about 10 billion years our aging sun will develop a core and an envelope that do not burn. There develops a hydrogen shell in the outer regions that burns and expands our aging sun into a RED GIANT. After another 2 billion years the red giant will have a radius of 1 A.U., and swallow the earth.

4. Galaxy Rotation and Dark Matter

GALAXY: A GARGANTUAN COLLECTION OF STARS, GAS, DUST, BLACK HOLES HELD TOGETHER BY GRAVITY AND ISOLATED IN SPACE.

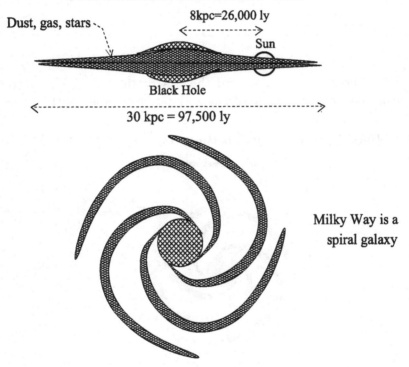

OUR GALAXY: THE MILKY WAY

Dust, gas, stars

8kpc=26,000 ly

Sun

Black Hole

30 kpc = 97,500 ly

Milky Way is a spiral galaxy

In 1932 the Dutch astronomer Jan Oort first discovered dark matter. This was only a few years after Hubble published his work on the velocities of galaxies using the Doppler shift. Jan Oort used the Doppler shifts of stars moving near the Milky Way galactic plane to determine their rotational velocities, and compared the centripetal acceleration to the gravitational acceleration of visible mass, like stars. He concluded that there must be about three times the amount of mass than visible mass. At about the same time Fritz Zwichy carried out studies of galactic clusters and found about 10 times the amount of mass as visible mass.

These studies, and many since then, measure rotation curves, as in the figure below for galaxy NGC3198.

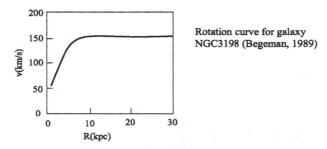

Rotation curve for galaxy NGC3198 (Begeman, 1989)

From the rotation curve and the observation of visible matter, one can determine the ratio of Dark Matter to visible matter.

4.1. *Milky Way Galaxy Rotation and Dark Matter*

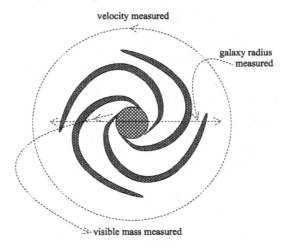

From the angular velocity and size of our galaxy, one can determine the mass. THERE IS MORE MASS THAN SEEN, THERE IS DARK MATTER. After years of study of galaxy formation, Cosmic Background Radiation, etc., we know that Dark Matter is not Standard Model matter. It is about 23% of matter in the universe, and will be discussed later.

4.2. *Active Galaxies, Quasars*

Active galaxies have greater luminosities than normal (spiral, elliptical) galaxies, and emit radiation with much longer wavelengths (mainly radio) rather than mainly visible radiation of normal galaxies.

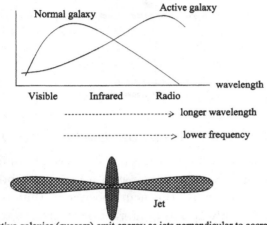

Active galaxies (quasars) emit energy as jets perpendicular to accretion disk, with the jets probably powered by large black holes.

QUASARS are a type of active galaxies detected at very large distances. They are the brightest objects that have been discovered in the universe.

5. Gravitational Collapse of a Massive Star

Massive Stars, $M > {\sim}8 \times M_{sun}$, burn their nuclear fuel quickly compared to our sun. After about one billion years or less they no longer have outward pressure from nuclear fusion to balance the gravitational pressure of their large mass.

They collapse to density $> 10^{14}\,\mathrm{g\,cm^{-3}} >$ nuclear density. Heavy elements form, making the secondary dust for gallaxy formation, as we discussed earlier.

Just after the collapse is the hydrodynamic phase, with what is called shocks, bounce, etc. See Ref. [2] for a discussion of the hydrodynamics needed for core collapse supernovae. In ${\sim}0.01\,\mathrm{s}$ a protostar is formed.

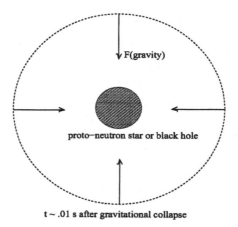

t ~ .01 s after gravitational collapse

5.1. *Neutrinosphere and Protostar During 1 s to 10 s*

About 1 s after the collapse of a large star thermal equilibrium is reached. At the center is a protostar, with a radius of about 10 km, surrounded by matter, called the neutrinosphere, that is so dense that neutrinos with only a weak interaction scatter after moving about 1 cm. That is, the neutrino's mean free path is about 1 cm. This is very important in understanding what produces the large velocity of rotating neutron stars, called pulsars.

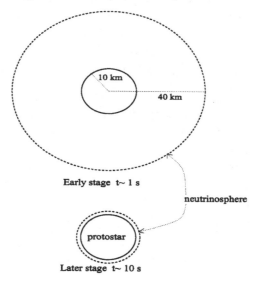

5.2. *The Final Fate of the Proto-Star: Pulsars and Black Holes*

After the interior star reaches thermal equilibrium, it is made of nuclear matter, largely neutrons, except in the interior where the nuclear matter is highly compressed.

The final star has a radius of about 10 km. If its mass is less than 1.5 M_{sun}, it becomes a neutron star, a pulsar. If its mass is greater than 1.5 M_{sun}, it continues to collapse to become a black hole.

The neutron star has a large magnetic field (arising from the original star's magnetic field), and spins very fast (conservation of angular momentum). This rapidly spinning magnetic field sends out electromagnetic radiation, which is seen on earth as short burst of light with a period of less than one second. Because of these rapid bursts of light, they are called pulsars.

Gravity keeps compressing a neutron-like star with a mass more than about 1 1/2 sun masses. When it reaches a critical radius, light is bent back by gravity, as predicted by Einstein's General Theory of Relativity, which we study in Chapter 6.

When radius of the proto-black hole reaches a critical radius, light emitted is bent back and is not visible. It is a black hole.

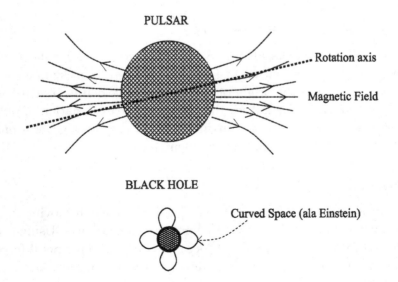

5.2.1. *Pulsar Velocities*

As stated above often the neutron star has a large magnetic field, is rotating rapidly, and since it emits rapid bursts of light it is called a pulsar. By the use of methods similar to Hubble's measurement of the velocity of distant galaxies, the speed of motion of the pulsars has been measured. Some pulsars move with very large velocities, with a mechanism called pulsar kicks.

5.2.2. *Supernovae*

After the collapse of a massive star and the dense matter around the star's remnants gets blown away, a very bright object called a supernova appears. As we mentioned above, this process of supernovae creation also leads to the creation of elements with atomic numbers larger than hydrogen and helium, the main constituents of primary cosmic gas from which the very early galaxies were formed. Without this secondary cosmic gas our planet earth would not have the atoms that were needed for our human civilization to evolve.

Also, supernovae were essential for determining the amount of Dark Energy in our present universe. We will discuss this in Chapter 7.

5.3. *Neutrinos and Pulsar Kicks*

This completes our chapter on Stars, Galaxies, Etc. In the next chapter we study the nature of neutrinos, the study of symmetries of our universe, and the origin of the very large velocities of neutron stars, pulsar kicks. We shall see that one possible explanation of pulsar kicks is the existence of neutrinos beyond the three generations in the standard model, called sterile neutrinos.

6. Problems

1. As was derived in Section 2, the speed of the earth moving in a circle with radius $R = 1.5 \times 10^{11}$ m is $v = 3 \times 10^4$ m/s. Using the mass of the earth $= 5.97 \times 10^{24}$ kg and equating centripetal force to gravitational force, find the force of gravity between the sun and the earth.

2a. Our sun is 26,000 ly from the center of our Milky Way galaxy. What is this distance in meters?

2b. If our sun is rotating about the center of our galaxy at 125 km/s, assuming that all the mass is in a very small volume at the center, what is the mass of our galaxy?

3. If during the first ten seconds after the collapse of a star neutrinos are emitted with a momentum of $p = 2 \times 10^{36}$ g km/s, what is the velocity of the neutron star, assuming that it has the mass of the sun?

4. During the ten seconds after the collapse of a massive star, with the neutrinosphere consisting mainly of neutrons, give a reaction that could have produced an anti-muon neutrino.

5. As discussed above, matter in the universe about 400,000 years after the Big Bang consisted mainly of hydrogen H and helium He^4. Give a possible set of nuclear reactions that could produce the lead atomic nucleus, Pb^{208}, with 82 protons and 126 neutrons.

References

[1] Eric Chaisson and Steve McMillan, *Astronomy Today*, published by Prentice-Hall, Inc.

[2] Anthony Mezzacappa and E.B. Messer, *J. Computational and Applied Mathematics*, **109**, 281 (1999).

Chapter 5

Neutrino Oscillations, Symmetries, and Pulsar Kicks

1. Overview of Chapter 5

In Chapter 5 we first discuss an important aspect of neutrinos, neutrino mass and neutrino flavor. Note that in our study of elementary particles we discussed the electron mass, but not the electron neutrino mass. This is because the electron neutrino has a flavor, the electron flavor, but does not have a well defined mass. This will lead to a number of very important and interesting topics.

(1) Energy and time dependence of quantum states: neutrino flavor states, neutrino mass states, and their time dependence.
(2) Neutrino oscillations.
(3) Parity, charge conjugation, and time reversal symmetries.
(4) Time reversal and CP symmetry violation and neutrino oscillations.
(5) Pulsar kicks with standard neutrinos.
(6) Pulsar kicks with sterile neutrinos.

2. Quantum States and Time Dependence of Neutrino States

Quantum theory is very complicated, and we do not treat it in this book. See, e.g., Liboff [1] for a detailed description of states, operators, expectation values and eigenvalues briefly discussed in the present section.

For the study of astrophysics a little quantum theory is needed. In particular we need to understand why neutrino states that were introduced in the first chapter, ν_e, ν_μ, ν_τ, are not stable in time. If one produces a beam of electron neutrinos and uses a neutrino detector at some distance from where the beam was created, one will detect both muon and tau neutrinos. This is called neutrino oscillations, an essential concept needed for the present chapter.

2.1. *Quantum Operators and Quantum States*

In quantum theory one does not deal directly with forces and the acceleration of masses, as in Newton's classical theory of motion. Instead one studies states and operators, from which one obtains information. We will not treat this rather complicated subject, but only what is needed to understand neutrino oscillation.

A quantum state represents some system that has a certain set of properties. Let us call these properties P for some particulate state:

State: $|P\rangle$

We are interested in the position and time dependence of state P, so let us represent the state by

$$|P(\vec{r}, t)\rangle$$

Quantum operators operate on quantum states, and in general change them into a different quantum state. For example the operator X

$$X|P\rangle = |Q\rangle, \quad \text{with the state } |Q\rangle \text{ different than the state } |P\rangle$$

For a particular state and operator, however, the operator does not change the state. For example the operator Z operating on the state $|z\rangle$

$$Z|z\rangle = z|z\rangle,$$

where z is a number. The state $|z\rangle$ is called an eigenstate of the operator Z, and z is the eigenvalue.

2.1.1. *Hamiltonian and Time Dependence of Energy Eigenstates*

Recall that in Chapter 1 we defined kinetic energy as the energy of motion and potential energy as stored energy, energy created by doing work on an object, such as lifting a mass from the floor to a table with no initial or final velocity. If heat energy is not involved, the total energy E is

$$E = \text{kinetic} + \text{potential energy}. \tag{1}$$

The Hamiltonian, H is the quantum operator of energy:

$$H = \text{quantum kinetic} + \text{potential energy}, \tag{2}$$

and the Hamiltonian operator also is the time derivative operator:

$$H = i(\partial/\partial t). \tag{3}$$

Since the Hamiltonian is the energy operator, a state with good energy is an eigenstate of the Hamiltonian:

$$H|A(\vec{r},t)\rangle = E_A|A(\vec{r},t)\rangle \quad \text{or} \quad i(\partial/\partial t)|A(\vec{r},t)\rangle = E_A|A(\vec{r},t)\rangle, \tag{4}$$

where we have used Eq. (3). From Eq. (4) one finds the time dependence of the energy eigenstate $|A(\vec{r},t)\rangle$

$$|A(\vec{r},t)\rangle = e^{-iE_A t}|A(\vec{r})\rangle, \tag{5}$$

since

$$i(\partial/\partial t)e^{-iE_A t} = E_A e^{-iE_A t}; \tag{6}$$

$$i(\partial/\partial t)|A(\vec{r},t)\rangle = E_A|A(\vec{r},t)\rangle. \tag{7}$$

We shall need this time dependence of a state with good energy to understand neutrino oscillations.

2.1.2. *Neutrino States With Good Mass and Time Dependence of Neutrino Flavor States*

We know that in the Standard Model there are three types of neutrinos, the electron, muon, and tau, which one calls flavor. There are also three mass neutrino eigenstates, with masses m_1, m_2, m_3. They are not the same states. The flavor states are a combination of the mass states. For example the electron neutrino

$$|\nu_e\rangle = c_1|m_1, t\rangle + c_2|m_2, t\rangle + c_3|m_3, t\rangle$$
$$|\nu_e\rangle = c_1 e^{-im_1 t}|m_1\rangle + c_2 e^{-im_2 t}|m_2\rangle$$
$$+ c_3 e^{-im_3 t}|m_3\rangle. \tag{8}$$

Since the three masses are different, the coefficients of an electron neutrino state change in time, and it therefore changes partly into a muon and a tau neutrino. This is called neutrino oscillation.

3. Three Important Symmetries in Quantum Theory: Parity, Charge Conjugation, and Time Reversal

In our study of neutrino oscillations to test symmetries, the three most important symmetries are parity, charge conjugation, and time reversal, with the quantum operators P, C, and T, respectively.

3.1. *Parity*

The parity operator, P, reverses direction, $(x, y, z) \rightarrow (-x, -y, -z)$. Therefore when it operates on a quantum state

$$P|\vec{r}, t\rangle = |-\vec{r}, t\rangle. \tag{9}$$

Particles have intrinsic parity. For example, pi mesons have negative parity, regardless of their electric charge

$$P|\pi^+\rangle = -|\pi^+\rangle$$
$$P|\pi^-\rangle = -|\pi^-\rangle \tag{10}$$
$$P|\pi^o\rangle = -|\pi^o\rangle,$$

while protons have positive parity

$$P|p\rangle = |p\rangle. \tag{11}$$

3.2. *Charge Conjugation*

The charge conjugation operator C changes a particle state to an antiparticle state. For fermions, like electrons or protons,

$$C|e^-\rangle = |e^+\rangle$$
$$C|p\rangle = |\bar{p}\rangle, \tag{12}$$

where $|e^+\rangle$ is a positron state, the positron being an anti-electron, while $|\bar{p}\rangle$ is an antiproton state. For mesons charge conjugation reverses the electric charge of the meson:

$$C|\pi^+\rangle = |\pi^-\rangle$$
$$C|\pi^o\rangle = |\pi^o\rangle. \tag{13}$$

3.3. *Time Reversal*

For time reversal we need both the time reversal operator T and its inverse $T^\dagger$, with

$$T^\dagger T = 1. \tag{14}$$

If the Hamiltonian is time-reversal invariant

$$THT^\dagger = H \text{ then}$$
$$Te^{-iH(t_2-t_1)}T^\dagger = e^{-iH(t_1-t_2)}. \tag{15}$$

3.4. *CP, T, and the CPT Theorem*

For any system which is defined by a Hamiltonian that is invariant to the transformations defined by Einstein's relativity theory, called Lorentz transformations, which is true for all systems with which we consider, the system is invariant to the combined operations CPT. This is called the CPT Theorem. Therefore if a system violates CP-invariance, it must also violate T-invariance.

3.5. *CP (and T) Violation in Weak Interactions*

In strong interactions CP, and therefore T has been found to be conserved, however in weak interactions it has long been known that

CP is violated. For example, the K^0 meson has strangness. That is, it consists of an s (strange) quark and an anti d quark (see Chapter 2). Pions have no strangeness. Only weak interactions can change strangness. Therefore when the K^0 meson decays to pions it is through the weak interactions.

It was observed many years ago that the K^0 mesons decay both into two π and three π states. From Eqs. (10), (13)

$$2 \text{ pion states have } CP = 1,$$

$$3 \text{ pion states have } CP = -1.$$

Therefore weak interactions violate *CP*.

In our studies of tests of symmetries via neutrino oscillations we discuss both *CP* and *T* violation, as experimental tests of these two symmetries are quite different, as we shall see.

4. Neutrino Flavor, Mass, and Oscillations

One mechanism for producing a beam of neutrinos is via proton–proton collisions. These collisions produce one of the three standard neutrino flavors ν_a, with $a = \text{flavor} = e, \mu, \tau$. E.g.,

$$p + p \rightarrow p + n + e^+ + \nu_e.$$

But as we discussed earlier, a neutrino with a flavor does not have a well-defined mass. Neutrinos of definite mass we label ν_α, with $\alpha = 1, 2, 3$. The ν_a, ν_α are 1 by 3 column matrices.

$$\nu_a = \begin{pmatrix} \nu_e \\ \nu_\mu \\ \nu_\tau \end{pmatrix} \qquad \nu_\alpha = \begin{pmatrix} \nu_1 \\ \nu_2 \\ \nu_3 \end{pmatrix}$$

flavor neutrinos   mass neutrinos

These two forms are connected by a 3×3 matrix [2]:

$$U = \begin{pmatrix} c_{12}c_{13} & s_{12}c_{13} & s_{13}e^{-i\delta_{CP}} \\ -s_{12}c_{23} - c_{12}s_{23}s_{13}e^{i\delta_{CP}} & c_{12}c_{23} - s_{12}s_{23}s_{13}e^{i\delta_{CP}} & s_{23}c_{13} \\ s_{12}s_{23} - c_{12}c_{23}s_{13}e^{i\delta_{CP}} & -c_{12}s_{23} - s_{12}c_{23}s_{13}e^{i\delta_{CP}} & c_{23}c_{13} \end{pmatrix},$$

with notation $s_{ij} = \sin\theta_{ij}$ and $c_{ij} = \cos\theta_{ij}$. The knowledge of the angles θ_{ij}, with $ij = 12, 13,$ and 23, are essential for tests of symmetry violations via neutrino oscillations, as we shall see. Flavor neutrinos are related to mass neutrinos by

$$\nu_a = U\nu_\alpha.$$

5. *T*, *CP* and *CPT* Violations in Neutrino Oscillations

As an introduction to the tests of symmetries via neutrino oscillations, we first define $P(\nu_\alpha \to \nu_\beta)$ = transition probability of flavor α to flavor β neutrino. The *T*, *CP*, and *CPT* violating probability differences are:

$$\Delta P_{\alpha\beta}^T = P(\nu_\alpha \to \nu_\beta) - P(\nu_\beta \to \nu_\alpha)$$
$$\Delta P_{\alpha\beta}^{CP} = P(\nu_\alpha \to \nu_\beta) - P(\bar{\nu}_\alpha \to \bar{\nu}_\beta)$$
$$\Delta P_{\alpha\beta}^{CPT} = P(\nu_\alpha \to \nu_\beta) - P(\bar{\nu}_\beta \to \bar{\nu}_\alpha),$$

where $\alpha, \beta = e, \mu, \tau$, are the three neutrino flavors. If an experiment detecting the flavor of neutrinos finds that $\Delta P_{e\mu}^T$ is not zero, this means that the electron to muon oscillation probability is not the same as the muon to electron probability, and thus there is *T*-violation. Similarly for $\Delta P_{e\mu}^{CP}$ and $\Delta P_{e\mu}^{CPT}$.

We use the time evolution matrix, $S(t, t_0)$ to derive $\Delta \mathcal{P}_{ab}^T, \Delta \mathcal{P}^{CP}$, and $\Delta \mathcal{P}^{CPT}$. See Ref. [3] for details on using $S(t, t_0)$ to derive *T* and *CP* violations via neutrino oscillations.

$$|\nu(t)\rangle = S(t, t_0)|\nu(t_0)\rangle \tag{16}$$

$$i\frac{d}{dt}S(t, t_0) = H(t)S(t, t_0), \tag{17}$$

with $H(t)$ the Hamiltonian, which includes the interaction of neutrinos with matter as the neutrino beam goes through earth. The potential, V, arising from neutrino scattering from electrons in the earth is

$$V = \sqrt{2}G_F n_e,$$

where G_F is the universal weak interaction Fermi constant, and n_e is the density of electrons in matter. Using the matter density $\rho = 3 \, \text{g/cm}^3$

$$V = 1.13 \times 10^{-13} \, \text{eV},$$

and $V(\text{antineutrinos}) = -V(\text{neutrinos}).$

The times in the S-matrix are t_0, the time at which the neutrino is created, as in the pp collision process defined in Section 4, and t, the time at which the neutrinos are detected. We take $t_0 = 0$. Since the neutrinos have a very small mass and have large energies compared to their mass, they travel almost with the speed of light $= c$. For convenience we use units with $c = 1$, so

$$t_0 = 0$$
$$t = L, \tag{18}$$

with L the distance from where the neutrinos are produced to where they are detected. We call L the baseline for the experiment.

Some neutrino oscillation experiments with flavor f1 oscillating to flavor f2 are:

t=0, produce neutrinos, flavor = f1   t, detect neutrinos

v_{f1} ---------------------------------------→ $v_{f1} + v_{f2}$

L=baseline. Since neutrino speed is almost that of light, c=1 in our units, L is approxiately=t=time at distance L.

E=energy of neutrino beam.

Some neutrino oscillation experiments, with baseline L and E = energy of neutrinos

	L	E
MINOS	735 km	3–18 GeV
MiniBooNE	500 m	0.5–1.5 GeV
JHF–Kamioka	395 km	0.4–1 GeV
CHOOZ	1 km	3 GeV

5.1. *Time Reversal Violation (TRV) for Electron–Muon Oscillations*

From Eq. (16) $\Delta P^T_{e\mu} = P(\nu_e \rightarrow \nu_\mu) - P(\nu_\mu \rightarrow \nu_e)$. One can show that the electron–muon transition probabilities in terms of the S-matrix elements are

$$P(\nu_e \rightarrow \nu_\mu) = S_{21}$$
$$P(\nu_\mu \rightarrow \nu_e) = S_{12} \quad \text{or} \tag{19}$$
$$\Delta P^T_{e\mu} = S_{21} - S_{12}.$$

For the angles in the U-matrix we use $s_{13} = 0.187$, and $s_{23} = c_{23} = 0.707$, $\theta_{12} = 32°$, and $\delta_{CP} = 90°$. Although neutrino masses are not known, we only need the mass differences: $\delta m_{31}^2 = 2.4 \times 10^{-3}$ eV^2, $\delta m_{21}^2 = 7.6 \times 10^{-5}$ eV$^2 \ll \delta m_{31}^2$. We also need the quantities $\delta = \delta m_{21}^2/(2E)$, $\Delta = \delta m_{13}^2/(2E)$, $2\omega = \sqrt{\delta^2 + V^2 - 2\delta V \cos(2\theta_{12})}$.

We do not give the details of the somewhat complicated derivation, but with these parameters one finds for the electron–muon time reversal violation (see Ref. [4] for details):

$$\Delta \mathcal{P}_{e\mu}^T \simeq 0.34 \sin \omega L \sin 2\theta (\cos \omega L - \cos \Delta L). \tag{20}$$

For neutrino energy $E = 1.0$ GeV we find $\Delta \mathcal{P}_{e\mu}^T$ for baseline lengths $L \leq 5000$ km, which includes MiniBooNE:

$$\Delta \mathcal{P}_{e\mu}^T \simeq 0.0032L(1.0 - \cos(34.4L)), \tag{21}$$

where L is in units of 1000 km. Using Eq. (21) we find the magnitude of the time reversal violation as a function of L shown in Fig. 5.1.

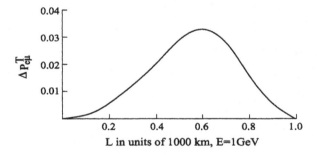

Fig. 5.1. $\Delta \mathcal{P}_{e\mu}^T$ with $E = 1$ GeV.

Next we treat the MINOS experiment, which has a baseline of $L = 735$ km. The energy range is 3–18 GeV. Minos cannot measure TRV, as it can only measure $\nu_\mu \rightarrow \nu_e$. We predict TRV in anticipation of future experiments. Using Eq. (20) one finds the result shown in Fig. 5.2.

Note that for low energies $\Delta \mathcal{P}_{e\mu}^T$ is 0.01 or more, which could be detected if a facility has both electron and muon neutrino beams.

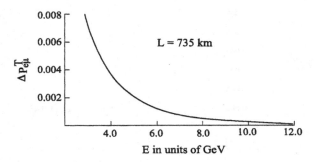

Fig. 5.2. $\Delta\mathcal{P}_{e\mu}^{T}$ for $L = 735$ km.

5.2.   *CP-Violation (CPV) for Electron–Muon Oscillations*

The CPV probability was recently estimated [5]. We now review CPV probability for $\Delta\mathcal{P}_{\mu e}^{CP} = \mathcal{P}(\nu_\mu \to \nu_e) - \mathcal{P}(\bar{\nu}_\mu \to \bar{\nu}_e) = |S_{12}|^2 - |\bar{S}_{12}|^2$ with S_{12} defined above, and $\bar{S}_{12} = S_{12}$ with neutrinos replaced by antineutrinos, with $V \to -V$. The result is shown in Fig. 5.3.

6.   Conclusions for Time Reversal and *CP*-Violation Tests

TRV tests require both $a \to b$ and $b \to a$, and CPV tests need beams of both neutrinos and antineutrinos. There is no facility at the present time that has these neutrino beams.

For TRV one possibility is using results from two different neutrino oscillation setups, one with a ν_e to measure $\nu_e \to \nu_\mu$ and the second with a ν_μ to measure $\nu_\mu \to \nu_e$.

Another proposed experiment to measure TRV is to put a second neutrino detector at distance $2L$. With the detector at baseline L, one can measure $\nu_e \to \nu_\mu$, and has a new beam of ν_μ neutrinos. Therefore with the detector at baseline $2L$ one can measure $\nu_\mu \to \nu_e$, and thereby measure TRV.

For CPV, since one needs both ν and $\bar{\nu}$ beams, the L, $2L$ method will not work. Therefore, with the present experimental neutrino beams, the best possibility for tests of CPV are to compare two different setups.

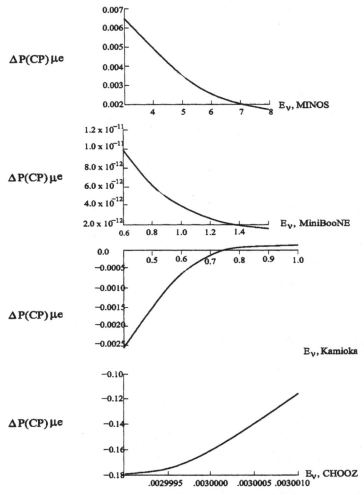

Fig. 5.3.   The ordinate is $\Delta\mathcal{P}(\nu_\mu \to \nu_e)$ for MINOS ($L = 735\,\mathrm{km}$), MiniBooNE ($L = 500\,\mathrm{m}$), JHF-Kamioka ($L = 295\,\mathrm{km}$), and CHOOZ ($L = 1\,\mathrm{km}$). Energy $= E$ in GeV.

7. Pulsar Kicks

As we briefly discussed in Chapter 2, the collapse of a massive star can create a black hole or a neutron star. Neutron stars have strong magnetic fields and rotate rapidly, sending out bursts of light. This is why they are called pulsars. See, e.g., Ref. [6].

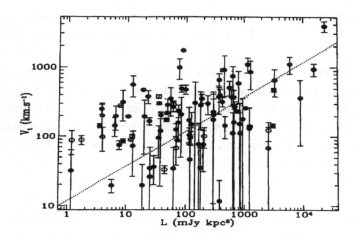

Using the Doppler shift, like Hubble, one can measure the velocity of pulsars, called pulsar kicks. It has been observed that pulsars with high luminosity, which also means high temperature, move with high velocity. The figure above shows the speed of pulsars as a function of the luminosity (brightness) of the pulsar.

The figure below illustrates the neutrinosphere, dense matter for which standard neutrinos have a mean free path of only a cm, during the first 10 seconds after the gravitational collapse.

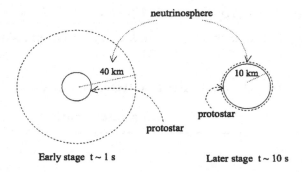

The time from the gravitational collapse to the formation of the what will become a pulsar is about 20 seconds. At about 0.1 second after the collapse the protoneutron star with a radius of about 10 km is surrounded by very dense matter, called the neutrinosphere, with a radius of about 40 km. After about 10 seconds the neutrinosphere's radius has reduced to about that of the protoneutron star, as illustrated above.

During the first 10 seconds, standard neutrinos produced by the URCA process cannot account for the pulsar kicks, while after 10 seconds the Modified URCA process can, as is discussed next.

From 0.1 to 10 s as the neutrinosphere contracts from $\sim$40 km to protostar radius $\sim$10 km neutrinos carry gravitational energy from the emerging star via URCA process

$$n \rightarrow p + e^- + \bar{\nu}_e$$

From $\sim$10 to $\sim$50 s $n-n$ collisions dominate neutrino production: star cooling via MODIFIED URCA process

$$n + n \rightarrow n + p + e^- + \bar{\nu}_e$$

Since standard neutrinos are scattered after travelling a very small distance in the neutrino sphere, they have a very small mean free path, the pulsar kicks cannot arise from active neutrinos during the first 10 s. As we discuss later, sterile neutrinos, which have no interaction, have a much longer mean free path than standard neutrinos.

7.1. *Pulsar Kicks from the Modified URCA Process*

For details on pulsar kicks from standard neutrino emission during the 10–50 s period see Ref. [7].

TYPICAL MODIFIED URCA DIAGRAM

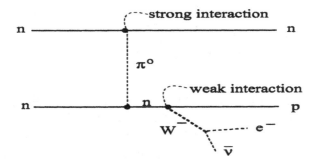

ELECTRONS IN n=0 LANDAU LEVEL:
NEUTRINOS EMITTED IN B−FIELD DIRECTION

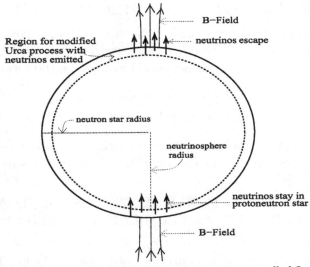

Electrons in a strong magnetic field are in quantum states, called Landau levels, similar to quantum levels in atoms, with n the principle quantum number. In the $n = 0$ level, the electron spin is only in the direction of the B-field, causing the neutrino created with the electron to move in the B-field direction.

The figure below shows the velocity in the T-range when the neutrinosphere is just inside the neutron star surface for electrons in $n = 0$ Landau level. Large pulsar velocities are found:

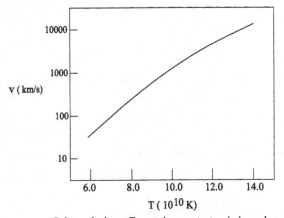

Pulsar velocity vs T assuming constant emission volume.

PULSAR KICKS FROM STERILE NEUTRINOS WITH LANDAU LEVELS IN A STRONG MAGNETIC FIELD (FIRST 10 SEC). STERILE NEUTRINOS HAVE NO INTERACTIONS.

Sterile/active neutrino mixing is given by mixing angle $|\theta_m|$. An electron neutrino can become a sterile neutrino, $|\nu_e\rangle \Rightarrow |\nu_s\rangle$. With the mixing angle θ_m, the two neutrino flavors have the form

$$|\nu_1\rangle = \cos\theta_m|\nu_e\rangle - \sin\theta_m|\nu_s\rangle$$
$$|\nu_2\rangle = \sin\theta_m|\nu_e\rangle + \cos\theta_m|\nu_s\rangle.$$

Strong magnetic fields result in neutrino asymmetries. Sterile neutrinos are emitted from the neutrinosphere in the direction of the magnetic field.

PULSAR KICK, TWO STERILE NEUTRINOS FOUND IN MINIBOONE EXPERIMENT AT FERMILAB, LSND EXPERIMENT AT LOS ALAMOS, ETC.

Pulsar kicks from sterile neutrinos with data on θ_m in 2009 were estimated in Ref. [8]. It was found that P_s = ratio of emission in direction of magnetic field to total emission:

$$P_s = 0.007-0.10, \tag{22}$$

which includes values of 1–2% needed to explain high velocity pulsar kicks. Because of uncertainty in θ_m, in 2009 specific predictions of pulsar velocity vs temperature were not made.

A new, much more accurate measurement of θ_m was published in 2012 [9]. The new value is:

$$\sin^2(2\theta_m) \simeq 0.15 \pm 0.05.$$

We recently made estimates of pulsar kicks using this new measurement: Kisslinger and Johnson [10].

Using for the momentum of the neutron star $p_{ns} = M_{ns}v_{ns}$ with the mass of the neutron star taken as $M_{ns} = M_{sun} = 2 \times 10^{33}$ g, one finds for the velocity of the neutron star as a function of temperature T, measured in units of 10^{10}K, with $\sin^2(2\theta_m) = 0.15$,

$$v_{ns} \simeq 22.3 \times 10^{-7} \left(\frac{T}{10^{10}\text{K}}\right)^7 \frac{\text{km}}{\text{s}}. \tag{23}$$

During the early stages after the collapse of a massive star temperatures $T = 20\,\text{MeV}$ are expected. With $T = 10$ to $20\,\text{MeV}$ the pulsar velocities, with a 50% range due to the uncertainty in $\sin^2(2\theta_m)$, are shown in the figure below.

Therefore, as shown in the figure, sterile neutrino emission can account for the large pulsar velocities for high luminosities (large T) that have been measured, as shown in the preceeding figure. This is a possible explanation of a puzzle that many have tried to explain for decades.

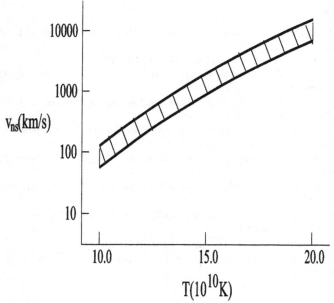

The pulsar velocity as a function of T for $\sin^2(2\theta) = 0.15 \pm 0.05$.

8. Problems

1. The operator for the x-component of momentum is $\hat{p}_x = -i(h/2\pi)\frac{d}{dx}$. Find the $\hat{p}_x$ eigenstate, $\hat{p}_x|p_x\rangle = p_x|p_x\rangle$.
2. Find the expectation value of $\hat{p}_x$ in the state $|p_x\rangle$.
3a. Starting from $\nu_a = U\nu_\alpha$ with a, α the neutrino flavor, mass, as 1×3 column vectors, show that $\nu_e = \sum_{j=1}^{3} U_{1j}\nu_j$.

3b. Using the formula in Problem 3a and the 3×3 matrix U given in the text, find ν_e in terms of ν_1, ν_2, ν_3 with the parameters c_{12}, etc.

4. The density of a neutron star is about $3 \times 10^{17}\,\mathrm{kg/m^3}$. What would your mass be if you were made of neutron star matter? How much would you weigh on the surface of the earth?

5. A 10 km radius neutron star pulsar is spinning 1000 times a second. What is the speed of a particle on the surface of the pulsar? Compare it to the speed of light.

References

[1] Richard L. Liboff, *Introductory Quantum Mechanics*, published by Addison Wesley.

[2] Jiro Arafune and Joe Sato, *Phys. Rev. D* **55**, 1653 (1997).

[3] Evgeny Akhmedov, Patrick Huber, Manfred Lindner and Tommy Ohlsson, *Nucl. Phys. B* **608**, 394 (2001).

[4] Ernest M. Henley, Mikkel B. Johnson and Leonard S. Kisslinger, *Int. J. Mod. Phys. E* **20**, 2463 (2011).

[5] Leonard S. Kisslinger, Ernest M. Henley and Mikkel B. Johnson, *Int. J. of Mod. Phys. E* **21**, 1250065 (2012).

[6] Eric Chaisson and Steve McMillan, *Astronomy Today*, published by Prentice-Hall, Inc.

[7] Ernest M. Henley, Mikkel B. Johnson and Leonard S. Kisslinger, *Phys. Rev. D* **76**, 125007 (2007).

[8] Leonard S. Kisslinger, Ernest M. Henley and Mikkel Johnson, *Mod. Phys. Lett. A* **24**, 2507 (2009).

[9] Kevork N. Abazajian *et al.*, arXiv: 1204.5379 (2012).

[10] Leonard S. Kisslinger and Mikkel B. Johnson, *Mod. Phys. Lett. A* **27**, 1250215 (2012).

Chapter 6

Einstein's Special and General Theories of Relativity

1. Overview of Chapter 6

In Chapter 6 we discuss both Einstein's Special Theory of Relativity and General Theory of Relativity. The Special Theory is a modification of Newton's theories, and the main difficulty in understanding this theory is that for objects moving very fast, the results differ from our intuitive concepts.

The General Theory of Relativity, which is essential for treating the evolution of the universe, is much more complicated. After an introduction to the metric tensor and other tensors needed to define Einstein's tensor, Einstein's equation of General Relativity is given.

Topics covered in the present chapter include:

(1) Einstein's Special Theory of Relativity: his postulates.
(2) Lorentz transformation.
(3) Length contraction and time dilation.
(4) Einstein's law of motion.
(5) Four-vectors and tensors.
(6) Ricci and energy-momentum tensors.
(7) Einstein's General Theory of Relativity.

2. Einstein's Special Theory of Relativity

Einstein's Special Theory of Relativity is needed for many aspects of astrophysics. For example (Chapter 3) the Doppler shift used by

Hubble to measure the expansion of the universe is based on relativity theory. Einstein's theory replaced Newton's Law of Motion by a law that has a similar form, but is quite different. See Pauli [1] for both the Special and General Theories of Relativity.

We begin this section with a brief review of Newton's law.

2.1. *Review of Velocity, Acceleration, Momentum, and Newton's Law of Motion*

First let us review the concept of acceleration $\vec{a}(t)$, a function of $t = $ time, with $\frac{d}{dt}$, $\frac{d^2}{dt^2}$ the first and second derivative with respect to time.

$$\text{Position} = \vec{r}(t)$$

$$\text{Velocity} = \vec{v}(t) = \frac{d\vec{r}(t)}{dt} = \dot{\vec{r}} \tag{1}$$

$$\text{Acceleration} = \frac{d\vec{v}(t)}{dt} = \frac{d^2\vec{r}(t)}{dt^2} = \ddot{\vec{r}}. \tag{2}$$

Newton's Law of Motion is that force = mass times acceleration:

$$\vec{F} = m \times \vec{a} = m \times \frac{d\vec{v}(t)}{dt}. \tag{3}$$

Next, momentum in Newton's theory is defined as

$$\text{Momentum} \equiv \vec{p} = m \times \vec{v}, \tag{4}$$

and therefore, using the fact that m is a constant,

$$\frac{d\vec{p}}{dt} = m \times \frac{d\vec{v}}{dt} = m \times \vec{a}. \tag{5}$$

Therefore Newton's Law of Motion can be written as

$$\vec{F} = \frac{d\vec{p}}{dt} = \dot{\vec{p}}. \tag{6}$$

As we shall see, Einstein's theory of motion looks similar, but is quite different.

Another very important difference between Einstein's and Newton's theories is the addition of velocities. In Newton's theory, if an object is moving with velocity $\vec{v}_1$ inside a rocket that is moving with velocity $\vec{v}_2$ with respect to earth, its velocity with respect to earth is

$$\vec{v}_1' = \vec{v}_1 + \vec{v}_2, \tag{7}$$

with the addition of vectors defined in Chapter 1. For example, if the direction of $\vec{v}_1$ is the same as $\vec{v}_2$, and the speed of the object with respect to the rocket and the speed of the rocket with respect to earch are both $0.8\,c$, then the speed of the object with respect to earth is

$$v_1' = (0.8 + 0.8) \times c = 1.6 \times c > c.$$

As we shall see, this is very different than the addition of velocities in Einstein's Special Theory of Relativity.

2.2. *Einstein's Postulates, Causality, and the Lorentz Transformation*

Einstein's two postulates needed for his Special Theory of Relativity are:

(1) **The laws of nature are the same in all inertial frames.**
(2) **The speed of light in vacuum measured in any inertial frame is equal to c, regardless of the motion of the light source with respect to that inertial reference frame.**

Causal Connection: These postulates give the result that information cannot be transmitted faster than the speed of light, which follows from the Lorentz transformation. See Einstein [2]. As we shall see in a later chapter this leads to the concepts of the radius of the universe, inflation and dark energy.

2.2.1. *Lorentz Transformation*

In 1904 Lorentz proved that Maxwell's equations, some of the most important equations in all of science, are invariant under what is now called the Lorentz transformation [3]. With the notation that (x, y, z) are the three axes and time $= t$ in the F frame, while (x', y', z') are the axes and time $= t'$ in the F' frame, which is moving in the

x-direction with speed v in the F frame, the Lorentz transformation is (with $\beta = v/c$ and $\kappa(v)$ a parameter to be determined)

$$x' = \kappa(v)\frac{x - vt}{\sqrt{1 - \beta^2}}$$

$$y' = \kappa(v)y$$

$$z' = \kappa(v)z \tag{8}$$

$$t' = \kappa(v)\frac{t - vx/c^2}{\sqrt{1 - \beta^2}}.$$

Now let us consider Lorentz transformation equations for a frame F'' which is moving parallel to the x-axis with speed $-v$ in the F' frame, with position, time coordinates (x'', y'', z'', t''). From Eq. (8)

$$x'' = \kappa(-v)\frac{x' + vt'}{\sqrt{1 - \beta^2}}$$

$$y'' = \kappa(-v)y'$$

$$z'' = \kappa(-v)z' \tag{9}$$

$$t'' = \kappa(-v)\frac{t' + vx'/c^2}{\sqrt{1 - \beta^2}}.$$

Since in the F'' frame the $+v$ in the F' frame has been reversed, so (x'', y'', z'', t'') is the same as (x, y, z, t), $x'' = \kappa(v)\kappa(-v)x$, $y'' = \kappa(v)\kappa(-v)y = y$, $z'' = \kappa(v)\kappa(-v)z = z$, $t'' = \kappa(v)\kappa(-v)t = t$. Therefore, since from symmetry $\kappa(v)$ must be independent of the direction of v, or $\kappa(-v) = \kappa(v)$, $\kappa(v) \times \kappa(-v) = \kappa(v)^2 = 1$, and therefore $\kappa(v) = 1$. Thus the Lorentz transformation, which is consistent with Einstein's Special Theory of Relativity, for v parallel to the x-axis is

$$x' = \frac{x - vt}{\sqrt{1 - \beta^2}}$$

$$y' = y \quad \text{and} \quad z' = z$$

$$t' = \frac{t - vx/c^2}{\sqrt{1 - \beta^2}} \tag{10}$$

$$x = \frac{x' + vt'}{\sqrt{1 - \beta^2}}$$

$$t = \frac{t' + vx'/c^2}{\sqrt{1 - \beta^2}}.$$

2.2.2. *Length Contraction and Time Dilation*

Consider a rod of length L_o in the F' frame at time $= t'$, which is moving with velocity v in the x-direction with respect to the F frame. The ends of the rod are $x_2'(t')$ and $x_1'(t')$, with

$$x_2'(t') - x_1'(t') = L_o. \tag{11}$$

In the rest frame F at time $= t$ the ends of the rod are $x_2(t)$ and $x_1(t)$, with length L

$$x_2(t) - x_1(t) = L, \tag{12}$$

as shown in the figure.

x'1 ▬▬▬▬▬▬▬▬▬▬▬▬▬▬▬▬▬▬▬▬▬▬▬▬ x'2

Length = L $_o$ (rod at rest)

x 1 ▬▬▬▬▬▬▬▬▬▬▬▬▬ x$_2$ - - - - - - - ⟶ v

Length = L (rod velocity=v)

From Eq. (10)

$$x_2' - x_1' = \frac{x_2 - vt}{\sqrt{1 - \beta^2}} - \frac{x_1 - vt}{\sqrt{1 - \beta^2}} = \frac{x_2 - x_1}{\sqrt{1 - \beta^2}} \text{ giving}$$

$$L = L_o\sqrt{1 - \beta^2}, \tag{13}$$

which is length contraction. The length of an object moving with velocity v is its length at rest times $\sqrt{1 - (v/c)^2} \leq 1$.

Time dilation is similar. If a clock is moving with velocity v with respect to a observer at rest, and the time interval of the moving clock is $\Delta t' = t_2' - t_1'$, then using Eq. (10), the time interval Δt measured by the observer at rest is

$$\Delta t = \frac{\Delta t'}{\sqrt{1 - \beta^2}}. \tag{14}$$

Therefore a clock in motion has a longer time interval than seen by an observer at rest.

This has led to what is called the twin paradox, discussed in Chapter 3, when we gave Eq. (14) with no proof. It is called the twin paradox as many people cannot believe that if Jim and Jane are twins and Jane travels with velocity .99c for 70 years as measured on earth, then Jim has aged 70 years while Jane has aged by only 10 years when she returns to earth, as you can prove using Eq. (14).

2.2.3. *Addition of Velocities and the Causal Connection*

An object is moving with velocity $v_1 = \frac{dx'}{dt'}$ in the x-direction on a train that is moving with velocity $v_2 = \frac{dx}{dt}$ in the same direction with respect to the ground. Then using the Lorentz Transformation, Eq. (10), the velocity of the object as measured by someone at rest on the ground, V, is

$$V = \frac{v_1 + v_2}{1 + (v_1 v_2)/c^2}. \tag{15}$$

For example, consider the addition of velocities with $v_1 = v_2 = 0.8c$. In Chapter 3, using Newton's theory we found $V = 1.6c$, while from Einsteins's theory, Eq. (15), $V = 0.976c$, less than the speed of light. From Eq. (15) the addition of any two velocities less that c will be less than c. Moreover from the square roots seen in the Lorentz Transformation $\sqrt{1 - (v/c)^2}$, Eq. (10), v must be less than c, since the square root of a negative number is imaginary. If $v > c$ then $1 - (v/c)^2 < 0$, and space and time coordinates would not be physical.

Thus from the Lorentz transformation the causal connection is proved. Information cannot be sent faster than the speed of light, one of the most important results from Einstein's Special Theory of Relativity.

2.2.4. *Einstein's vs Newton's Laws of Motion*

In the Special Theory of Relativity, momentum is defined as

$$\vec{p} \equiv \frac{m\vec{v}}{\sqrt{1 - v^2/c^2}}, \tag{16}$$

while for Newton $\vec{p} = m\vec{v}$. In both theories the law of motion is

$$\vec{F} = \frac{d\vec{p}}{dt}, \tag{17}$$

with the force, $\vec{F}$, the same in both theories, but with the two different definitions of momentum, $\vec{p}$.

3. Einstein's General Theory of Relativity

See Weinberg [4] for a detailed treatment of tensors and the General Theory of Relativity.

As we discussed in Chapter 1, the most important force for the evolution of the universe is the gravitational force. The standard model treats the strong, electromagnetic, and weak forces, but not the gravity, since quantum field does not work for gravity.

In the second decade of the 20th century, Einstein expanded his Special Theory of Relativity to include gravity. Although we will not need all the details given in this section, we present a brief treatment of the General Theory introducing four-vectors and tensors; and shall derive a simple form in the following chapter.

3.1. *Four-Vectors*

Einstein's General Theory of Relativity make use of four-vectors and tensors. First we review three-vectors, like position $= \vec{r}$.

3.1.1. *Review of Three-Vectors*

It is convenient to represent a vector in terms of unit vectors, $\hat{e}_i$, which has a unit length and points in the i-direction. The notation $i = (1, 2, 3) = (x, y, z)$ is used for taking sums over the vector indices. An example is the position vector, $\vec{r}$, shown in the figure below.

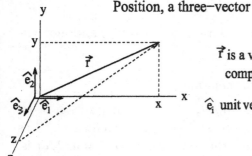

Position, a three–vector

$\vec{r}$ is a vector, with x, y, and z the three components.

$\hat{e}_i$ unit vector : i = (1,2,3) for (x,y,z) directions

$$\vec{r} = x\hat{e}_1 + y\hat{e}_2 + z\hat{e}_3 = \sum_{i=1}^{3} x^i\hat{e}_i \equiv x^i\hat{e}_i, \tag{18}$$

with the sum from 1 to 3 implied.

3.1.2. *Scalar Product of Three-Vectors*

First, the scalar product of unit vectors is

$$\begin{cases} \hat{e}_1 \cdot \hat{e}_1 = \hat{e}_2 \cdot \hat{e}_2 = \hat{e}_3 \cdot \hat{e}_3 = 1, \\ \hat{e}_i \cdot \hat{e}_j = 0 \quad \text{if } i \neq j, \end{cases}$$

or

$$\hat{e}_i \cdot \hat{e}_j = \delta_{ij}, \tag{19}$$

with δ_{ij} equals 1 if i = j and 0 if not. From Eq. (19) the scalar product of two three-vectors, $\vec{A} = A^i \hat{e}_i$ and $\vec{B} = B^i \hat{e}_i$ is

$$\vec{A} \cdot \vec{B} = \sum_{i=1}^{3} A^i B^i \equiv A^i B^i. \tag{20}$$

3.1.3. *Definition of Four-Vectors*

A four vector, $\vec{A}$, has four components, $\vec{A} = (A^0, A^1, A^2, A^3)$, or

$$\vec{A} = A^0 \hat{e}_0 + A^1 \hat{e}_1 + A^2 \hat{e}_2 + A^3 \hat{e}_3$$
$$= A^\alpha \hat{e}_\alpha \quad \text{with } \alpha = 0, 1, 2, 3. \tag{21}$$

As examples the space-time and momentum four-vectors are

$$\vec{r} = (x^0 = ct, x^1, x^2, x^3)$$
$$\vec{p} = (E/c, p^1, p^2, p^3), \tag{22}$$

where the $(1, 2, 3)$ components are those of the position and momentum three-vectors.

3.1.4. *Four-Velocity and Four-Momentum*

Since the definition of velocity is $d\vec{r}/dt$, and the four-position of an object moving with velocity v in the x-direction is given as x^i, x^0 as a function of time $= t'$ in Eq. (10), the four-velocity $U^\alpha = dx^\alpha/dt$ is

$$U^0 = \frac{c}{\sqrt{1 - (v/c)^2}}$$

$$U^1 = \frac{v}{\sqrt{1 - (v/c)^2}} \tag{23}$$

$$U^2 = U^3 = 0.$$

From Eq. (16) the relativistic four-momentum p^α for a particle of mass m moving in the x-direction with three-velocity v and four-velocity U^α, with energy $E = p^0 c$ is (with $p^2 = p^3 = 0$)

$$p^x = p^1 = \frac{mv}{\sqrt{1 - (v/c)^2}}$$

$$E = p^0 \times c = \sqrt{p^2 c^2 + m^2 c^4}. \tag{24}$$

Note that from $E = p^0 c$ and $p^0 = mU^0$, from Eq. (23) the well-known expression for E given in Eq. (24) is easily derived. For p $= 0$ this gives the famous $E = mc^2$.

3.1.5. *Scalar Product of Four-Vectors and Lorentz Invariance*

The scalar product of two four-vectors, $\vec{A}$, $\vec{B}$ is

$$\vec{A} \cdot \vec{B} = \sum_{i=1}^{3} A^i B^i - A^0 B^0 \tag{25}$$

$$\vec{r} \cdot \vec{r} = r^2 - (ct)^2$$

$$\vec{p} \cdot \vec{p} = p^2 - (E/c)^2,$$

where $r^2 = x^2 + y^2 + z^2$ and $p^2 = p_x^2 + p_y^2 + p_z^2$.

The scalar product of two four-vectors is invariant to a Lorentz transformation, $A' \cdot B' = A \cdot B$. Therefore from Eq. (25)

$$r'^2 - c^2 t'^2 = r^2 - c^2 t^2$$

$$p'^2 - (E'/c)^2 = p^2 - (E/c)^2. \tag{26}$$

3.2. **Tensors**

The tensors with which we deal in four-dimensional space-time are 4×4 quantities which transform from one frame to another with a Lorentz transformation. Defining the Lorentz transformation from one inertial frame to another in terms of the space-time four-vector $x^\mu = (x^0 = ct, x^1, x^2, x^3)$

$$x'_\mu = \sum_{\nu=0}^{3} \alpha_{\mu\nu} x_\nu, \tag{27}$$

a tensor t_{rs} transforms as

$$t'_{rs} = \sum_{k=0}^{3} \sum_{l=0}^{3} \alpha_{rk} \alpha_{sl} t_{kl}. \tag{28}$$

3.2.1. *Metric Tensor in Curved Space*

The standard notation for $\vec{r} \cdot \vec{r} = r^2 - (ct)^2$ is $s^2 = r^2 - (ct)^2$ in flat space. The definition of the metric tensor, $g_{\mu\nu}$ for flat space is

$$s^2 = g_{\mu\nu} x^\mu x^\nu = x^2 + y^2 + z^2 - c^2 t^2$$

$$g_{00} = -1$$

$$g_{ii} = 1 \tag{29}$$

$$g_{\mu\nu} = 0 \quad \text{if } \mu \neq \nu$$

In the general theory of relativity, where space can be curved, the metric tensor is a function of space-time $g_{\mu\nu} = g_{\mu\nu}(x^\alpha)$

$$s^2 = g_{\mu\nu}(x) x^\mu x^\nu \tag{30}$$

3.2.2. *Riemann Curvature Tensor and Ricci Tensor*

Einstein's General Theory predicts that space is curved near a large mass, and a beam of light does not travel in a straight line when it passes through. For the equations of general relativity we need the Riemann and Ricci tensors.

Assuming symmetry, $g_{\mu\nu}(x) = g_{\nu\mu}(x)$, and using $g_{\mu\nu}(x) = g^{\nu\mu}(x)$, the Riemann tensor, $\mathcal{R}^\alpha_{\beta\mu\nu}$, is

$$\mathcal{R}^\alpha_{\beta\mu\nu} = \frac{1}{2} \sum_\lambda g^{\alpha\lambda} (g_{\lambda\nu,\beta\mu} - g_{\lambda\mu,\beta\nu} + g_{\beta\mu,\lambda\nu} - g_{\beta\nu,\lambda\mu}) \tag{31}$$

$$g_{\lambda\nu,\beta\mu} \equiv \partial_\beta \partial_\mu g_{\lambda\nu}.$$

The Ricci tensor, $\mathcal{R}_{\mu\nu}$, is obtained from the Riemann tensor, $\mathcal{R}^\alpha_{\mu\alpha\nu}$, with a sum over α

$$\mathcal{R}_{\mu\nu} = \sum_\alpha \mathcal{R}^\alpha_{\mu\alpha\nu}. \tag{32}$$

3.2.3. *Homogeneous Space and the Robertson–Walker Metric*

As we shall see from the Cosmic Microwave Background measurements in Chapter 8, the universe is homogeneous except for galaxies and other astronomical objects distributed in space, and very small effects from early universe events. It is convenient to use spherical coordinates, shown in the figure, for the metric tensor.

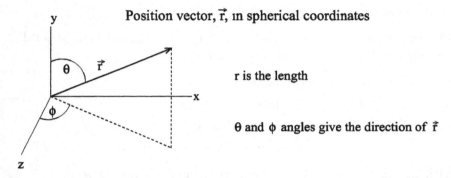

Position vector, $\vec{r}$, in spherical coordinates

r is the length

θ and φ angles give the direction of $\vec{r}$

    Using the coordinates r, θ, ϕ shown in the figure, the Robinson–Walker metric is

$$ds^2 = R^2(t)\left[\frac{dr^2}{1-kr^2} + r^2 d\theta^2 + r^2 \sin^2\theta d\phi^2\right], \qquad (33)$$

with $R(t)$ the cosmic scale factor, to be determined. The quantity k represents the curvature of spaces, with $k = 1, -1, 0$ for positive, negative, or zero spatial curvature. We discuss curvature in the next chapter.

    The Ricci tensor has a simple form for the homogeneous universe:

$$\mathcal{R}_{00} = 3\frac{\ddot{R}}{R}$$

$$\mathcal{R}_{ij} = \left[\frac{\ddot{R}}{R} + 2\frac{\dot{R}^2}{R^2} + \frac{2k}{R^2}\right]g_{ij}, \qquad (34)$$

and the Ricci scalar, $\mathcal{R}$, defined as

$$\mathcal{R} = \sum_{\mu=0}^{3}\sum_{\nu=0}^{3} g^{\mu\nu}\mathcal{R}_{\mu\nu}, \qquad (35)$$

is

$$\mathcal{R} = 6 \left[\frac{\ddot{R}}{R} + \frac{\dot{R}^2}{R^2} + \frac{k}{R^2} \right]. \tag{36}$$

We shall see that with a homogeneous universe, Einstein's equations of general relativity take a rather simple form.

3.2.4. *Energy-Momentum Tensor*

The last ingredient that we need for Einstein's equations of general relativity is the energy-momentum tensor, $T^{\alpha\beta}$,

$$T^{00} = \text{energy density}$$
$$T^{0i} = \text{energy flux across } i \text{ surface}$$
$$T^{i0} = i \text{ momentum density}$$
$$T^{ij} = i \text{ momentum flux across } j \text{ surface}, \tag{37}$$

with flux being the amount of energy or momentum per second.

In the frame with $U^0 = \frac{c}{\sqrt{1-(v/c)^2}}, U^1 = \frac{v}{\sqrt{1-(v/c)^2}}, U^2 = U^3 = 0$ the energy-momentum tensor is

$$T^{00} = \frac{\rho}{\sqrt{1 - (v/c)^2}}$$

$$T^{0i} = \frac{\rho v^i}{\sqrt{1 - (v/c)^2}} = T^{i0}$$

$$T^{ij} = \frac{\rho v^i v^j}{\sqrt{1 - (v/c)^2}}. \tag{38}$$

This is what one would expect for energy density, T^{00} ($=\rho$ for a system at rest, $v = 0$); and the energy flux ($T^{0i} = v^i T^{00}$), and momentum flux ($T^{ij} = v^j T^{0i}$) with $v^i = (v^x, v^y, v^z) = (v, 0, 0)$.

3.3. *Einstein's Equation with the Cosmological Constant*

The Einstein tensor $G^{\alpha\beta}$ is defined in terms of the Ricci tensor and scalar as

$$G^{\alpha\beta} = \mathcal{R}^{\alpha\beta} - \frac{1}{2}\mathcal{R}g^{\alpha\beta}, \tag{39}$$

and Einstein's equation of general relativity is

$$G^{\alpha\beta} = \mathcal{R}^{\alpha\beta} - \frac{1}{2}\mathcal{R}g^{\alpha\beta} = 8\pi G T^{\alpha\beta} + \Lambda g^{\alpha\beta}, \qquad (40)$$

where G is Newton's gravitational constant and Λ is the cosmological constant, which we discuss in the next chapter.

Einstein introduced the cosmological constant after other theorists showed that his original equation, Eq. (40) without the $\Lambda g^{\alpha\beta}$ term, had no static solution. That is, the universe was predicted to either expand or contract. After Hubble showed that the universe is expanding, Einstein removed this Λ term, and it is often stated that Einstein said that it "was my greatest blunder" to insert the cosmological constant. As we discuss in the next chapter, the cosmological constant was reintroduced recently to solve what is called the horizon problem. Λ is a simple mechanism for introducing dark energy, the energy of the vacuum, which is a large part of the content of the universe. This is discussed briefly in the following section and discussed in detail in several of the following chapters.

The equation given in Eq. (40) involves relations between tensors, and is very complicated. However, it is one of the most important equations in all of science. From it one can derive the size and temperature of the universe, which predicts important cosmological events, such as cosmological phase transitions when particles got mass and the universe went from a dense fluid of quarks and gluons to our present universe with proton and neutrons. This will be treated in detail in the next chapter.

4. Einstein's General Theory of Relativity and the Evolution of the Universe: Discussed in Chapter 7

In the next chapter we first derive a much simpler form of Einstein's equation for a homogeneous universe to get the Friedmann equation. From this we discuss tests of the the General Theory of Relativity: the bending of a light beam as it goes by a large mass, like the sun, and the gravitational Doppler shift. Gravitational radiation, which is related to the gravitational Doppler shift, is discussed. From gravitational radiation a great deal of information about the early universe could be found, but at the present time it has not been detected.

Then the size $R(t)$ and temperature $T(t)$ of the universe as a function of time are derived, from which one understands the evolution of the universe. We also discuss and derive inflation. At a very early time t_o there is a problem as $R(t_o)$ is larger than the causally connected distance at time t_o. Thus as our universe evolved and $R(t)$ grew, our present universe would be made of many different universes, and would not be uniform. By inserting the cosmological constant Λ in Friedmann's equation, one gets a causally connected, homogeneous universe. This is derived and discussed in detail in the next chapter. The cosmological constant, Λ, represents Dark Energy. The experimental detection of Dark Energy, which might not be correctly represented by Λ, is discussed in later chapters.

5. Problems

1a. The F' frame is moving with velocity $v = 0.1\,c$ in the x-direction. If $x = y = z = 0$ and $t = 0$, what are x', y', z', and t'?

1b. Same as 1a. except $t = 1\,\text{s}$ ($x = y = z = 0$). What are x', y', z', t'?

2. A rocket $25\,\text{m}$ long is moving with speed $v = 0.9c$ with respect to the ground, and a clock in the rocket says $\Delta t' = 30\,\text{s}$, what does an observer on the ground measure as the length of the rocket and the time interval?

3. A car in the rocket of Problem 2 is moving with speed $v = 0.9c$ with respect to the rocket. What does an observer on the ground measure as the speed of the car?

4. With the four-velocity U^α and p^α defined in Eqs. (23) and (24), with $\alpha = 0, 1, 2, 3$, what is the scalar product of these two four-vectors, $\vec{U} \cdot \vec{p}$?

References

[1] Wolfgang Pauli, *Theory of Relativity*, published by Dover Publications, Inc.
[2] Albert Einstein, *Ann. Phys.* **23**, 71 (1907).
[3] Hendrik A. Lorentz, *Proc. Acad. Sci. Amst.* **12**, 986 (1904).
[4] Steven Weinberg, *Gravitation and Cosmology: Principles and Applications of the General Theory of Relativity*, published by John Wiley & Sons, Inc. (1972).

Chapter 7

Radius and Temperature
of the Universe from the General
Theory of Relativity

1. Overview of Chapter 7

In the present chapter a simplified form of Einstein's equation of General Relativity is presented — Friedmann's equation — with a number of applications. The topics in Chapter 7 include

(1) Tests of Special and General Relativity.
(2) Black holes.
(3) Friedmann's equation for the radius of the universe, $R(t)$.
(4) Solutions for $R(t)$ using thermodynamics.
(5) Solution of Friedmann's equation for temperature $T(t)$.
(6) The horizon problem: inflation with the cosmological constant.
(7) The horizon problem: inflation with the quintessence field.
(8) Gravitational waves and quantum gravity.
(9) Einstein vs Bohr on relativity and quantum mechanics.

2. Tests of Theories of General
and Special Relativity

We begin with a brief review of Einstein's equations of General and Special Relativity.

2.1. *Review of Equations of General and Special Relativity*

From Chapter 6, Einstein's equation of General Relativity without the cosmological constant is

$$\mathcal{R}^{\alpha\beta} - \frac{1}{2}\mathcal{R}g^{\alpha\beta} = 8\pi G T^{\alpha\beta}, \tag{1}$$

where $\mathcal{R}^{\alpha\beta}$ is the Ricci tensor, $\mathcal{R}$ is the Ricci scalar, G is Newton's gravitational constant, and $T^{\alpha\beta}$ is the energy-momentum tensor. See Roos [1] for Special Relativity and tensors needed for the General Theory of Relativity.

In the present chapter symmetry will be used to simplify these tensors and derive a much simpler equation for determining the radius and temperature of the universe at any time. For the evolution of the universe one uses $T^{\alpha\beta}$ for the universe at any time, which as we see will be given by the energy density and pressure of the universe.

We need the momentum $\vec{p}$ and the energy E of an object with mass M and velocity $\vec{v}$. From Einstein's Special Theory of Relativity

$$\vec{p} = M\vec{v}/\sqrt{1 - v^2/c^c}$$
$$E = \sqrt{p^2c^2 + M^2c^4}. \tag{2}$$

Two other main results of the Special Relativity, in addition to the limit of speed to the speed of light $= c$ given by the Lorentz transformation, are length contraction and time dilatation. If a rod having a length L_o when at rest is moving with speed v with respect to the ground in the direction of the rod, its length L measured by an observer on the ground is

$$L = L_o\sqrt{1 - (v/c)^2}, \tag{3}$$

with $L < L_o$ as $\sqrt{1 - (v/c)^2} < 1$ for finite v (which must be less than c). Thus the length of a moving object is contracted.

If a clock is moving with velocity v with respect to a observer at rest on the ground, the time interval Δt measured by the observer at rest is related to the time interval $\Delta t'$ measured by the moving clock by

$$\Delta t = \frac{\Delta t'}{\sqrt{1 - (v/c)^2}}, \tag{4}$$

so the time measured by the observer on the ground for $v \neq 0$ is longer than that measured by the moving clock, called time dilatation.

We now discuss the tests of both Einstein's Special and General Theories.

2.2. Tests of the Special Theory of Relativity

First, the fact that the energy of a particle with mass M and momentum p is $\sqrt{p^2c^2 + M^2c^4}$ has been proved to be correct many times. For example, if one produces a particle of mass M moving with momentum $\vec{p}$ in an accelerator and it decays to two particles with mass m_1 and m_2 with measured momenta $\vec{p}_1$ and $\vec{p}_2$, illustrated in the figure below,

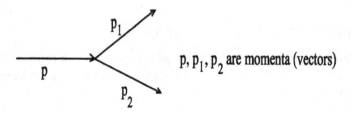

one finds

$$\vec{p} = \vec{p}_1 + \vec{p}_2, \tag{5}$$

which is conservation of momentum and

$$\sqrt{p^2c^2 + M^2c^4} = \sqrt{p_1^2c^2 + m_1^2c^4} + \sqrt{p_2^2c^2 + m_2^2c^4}, \tag{6}$$

which is conservation of energy with $E = \sqrt{p^2c^2 + M^2c^4}$ from Einstein's Special Theory of Relativity within the error of the experiment. Also from Einstein's Special Theory, one of the most striking results is time dilation, Eq. (4). This has also been tested many times via particle accelerators. Using the fact that a particle with velocity

v travels a distance $v\Delta t$ in time Δt, by measuring the distance that a particle with known lifetime travels with various speeds before it decays, Einstein's time dilatation has been proved to be correct many times.

2.3. *Tests of the General Theory of Relativity*

Einstein's General Theory of Relativity included gravity in his theory of relativity. Although gravity is a very weak force in the sense that it is neglected in comparison to the electric force when treating atoms or even the weak force when studing neutrinos, the force of gravity is important for large masses like the sun, or in the early universe when matter is very dense. Therefore the tests of the General Theory of Relativity are measuring the effects of large masses. We treat two such studies here, the gravitational shift in the frequency of light emitted from an atom, which we call the gravitational Doppler shift, and the change in the direction of a beam of light as it passes the sun.

2.3.1. *Gravitational Doppler Shift of Spectral Lines*

Recall from our discussion Bohr's model of atoms in Chapter 2, that the frequency, ν of the photon, the quantum of the elecromagnetic field, is given by the relationship

$$h\nu = E_i - E_f, \tag{7}$$

where E_i and E_f are the initial and final atomic energies and h is Planck's constant.

Consider an atom near the surface of our sun, which has a large mass, M_{sun}. Calling the radius of the sun r, the atom feels a gravitational force, $F_g(r)$, that is proportional to mass divided by the square of radius in the $-r$ direction:

$$F_g(r) \propto -\frac{M_{sun}}{r^2}. \tag{8}$$

Recall from Chapter 1, force is obtained from the potential, $V(r)$ by $F(r) = -\frac{dV(r)}{dr}$, so using $\frac{d(1/r)}{dr} = -\frac{1}{r^2}$ the gravitational potential felt by the atoms near the surface of the mass is [2]

$$V_g(r) \propto -\frac{M_{sun}}{r}.$$ (9)

This produces a frequency shift due to the difference in the energy of the atom near the sun compared to that on earth, which is far from the sun and feels a much smaller gravitational potential, since the mass of the earth is very much smaller than that of the sun. After including the various constants, and neglecting the gravitational effect on earth, this leads to a frequency shift $\Delta\nu$,

$$\frac{\Delta\nu}{\nu} \simeq -\frac{M_{sun}}{rc^2},$$ (10)

which is an important prediction of the General Theory of Relativity. Using $M_{sun} = 1.983 \times 10^{33}$ g and the radius of the sun $r = 6.95 \times 10^{10}$ cm, from Eq. (10) one finds

$$\frac{\Delta\nu}{\nu} \simeq -2.12 \times 10^{-6}.$$ (11)

This small red-shift was confirmed soon after Einstein's 1911 prediction. The observed frequency is not that predicted by Eq. (11), however, because of the motion of the atoms on the sun. Since the sun is very hot, the atoms have a rather large velocity, and the Doppler shift due to this motion changes $\frac{\Delta\nu}{\nu}$. When this is taken into account Einstein's prediction is verified.

2.3.2. *The Gravitational Deflection of Light*

When a beam of light passes a mass it feels a gravitational field that deflects the direction of the beam, as shown in the figure.

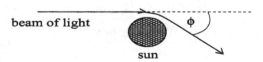

Using Einstein's equations, the angle of deflection, ϕ, is obtained with a method similar to that for finding the gravitational Doppler shift. The details can be found in Ref. [2]. The angle ϕ for a beam of light passing near the sun, as observed on earth is

$$\phi = 4 \times \frac{M_{sun}}{rc^2}$$

$$= 8.45 \times 10^{-6} \text{ radians.} \tag{12}$$

This very small angle is difficult to measure, but it is in agreement with observations of stars whose light passes close to the sun during a total eclipse of the sun.

Therefore both Einstein's Special and General Theories of Relativity have been confirmed by experimental observations.

2.3.3. *Schwarzschild Radius and Black Holes*

Recall that the metric for flat space is $s^2 = r^2 - c^2t^2$, or in spherical coordinates (see the figure before Eq. (33) in Chapter 6)

$$ds^2 = dr^2 + r^2(d\theta^2 + \sin^2\theta d\phi^2) - c^2dt^2. \tag{13}$$

In 1916, K. Schwarzschild found a solution for ds^2 consistent with Einstein's equation, Eq. (1), in the region of a mass $= M$,

$$ds^2 = \frac{dr^2}{1 - \frac{2GM}{c^2r}} + r^2(d\theta^2 + sin^2\theta d\phi^2) - \left(1 - \frac{2GM}{c^2r}\right)dt^2. \tag{14}$$

From Eq. (14) there is a singularity at a distance $r_{Sch} =$ Schwarzschild radius from the center of the mass (See Roos [1])

$$r_{Sch} = \frac{2GM}{c^2}, \tag{15}$$

where G is Newton's gravitational constant. If $r_{Sch} > r_M$, with r_M the radius of the mass, then light leaving the mass is deflected so much that it curves around and is not emitted. The mass is a black hole. For example, for $M = 10 \times M_{sun} = 1.99 \times 10^{31}$ kg,

$$r_{Sch} = \frac{2 \times 6.67 \times 10^{-11} \times 1.99 \times 10^{31}}{(3.0 \times 10^8)^2} m = 30 \, \text{km.} \tag{16}$$

Note that the Schwarzschild radius is proportional to the mass, so for the sun $r_{Sch} = 3$ km, which is within the sun. Therefore Eq. (15) cannot be used for our sun, which is not a black hole.

Black holes are formed during the collapse of a massive star, and are often at the center of galaxies, as discussed in Chapter 4.

3. Friedman's Equations for $R(t)$ = Radius of the Universe

We now derive Einstein's equation of General Relativity without the cosmological constant, Eq. (1), for a homogeneous isotropic universe. See Kolb and Turner [3] for a detailed derivation and solution to Friedmann's equations.

From symmetry the energy-momentum tensor must be diagonal, $T_{\alpha\beta} = 0$ if $\alpha \neq \beta$. From Eq. (38), Chapter 6, for a perfect fluid with an energy density $\rho(t)$ and pressure $p(t)$,

$$
\begin{aligned}
T_{00} &= -\rho(t) \\
T_{ii} &= p(t) \\
\rho &= \text{energy/volume} \\
p &= \text{pressure} = \text{force/area}.
\end{aligned}
\tag{17}
$$

The Ricci tensor is diagonal. Defining $R(t)$ the radius of the universe for time $= t$, the Ricci tensor $\mathcal{R}_{\alpha\beta}$ and Ricci scalar $\mathcal{R}$ are

$$
\mathcal{R}_{00} = -3\frac{\ddot{R}(t)}{R(t)}
$$

$$
\mathcal{R}_{ii} = -\left(\frac{\ddot{R}(t)}{R(t)} + 2\frac{(\dot{R}(t))^2}{R(t)^2} + \frac{2k}{R(t)^2} \right)
\tag{18}
$$

$$
\mathcal{R} = -6\left(\frac{\ddot{R}(t)}{R(t)} + \frac{(\dot{R}(t))^2}{R(t)^2} + \frac{k}{R(t)^2} \right).
$$

From this the 00 component of Eq. (1), $\mathcal{R}_{00} - \mathcal{R}/2 = 8\pi G T_{00}$ is

$$
\frac{\dot{R}^2}{R^2} + \frac{k}{R^2} = \frac{8\pi G}{3}\rho,
\tag{19}
$$

which is often called the Friedmann's equation, with k the curvature constant, discussed below. The ii component gives

$$
2\frac{\ddot{R}}{R} + \frac{\dot{R}^2}{R^2} + \frac{k}{R^2} = -8\pi G p.
\tag{20}
$$

Subtracting Eq. (19) from Eq. (20) one obtains the important equation for the acceleration of $R(T)$, Friedmann's second equation.

$$\frac{\ddot{R}(t)}{R(t)} = -\frac{4\pi G}{3}(\rho + 3p). \tag{21}$$

3.1. *Units of Pressure and* ρ

Note that p = pressure = force/area = F/A. Volume of an object with an area A and width d is $V = A \times d$, or $A = V/d$. Therefore

$$p = F/A = \frac{F \times d}{V} = \frac{\text{Work} = \text{Energy}}{V}, \tag{22}$$

while pressure = $p = \frac{E}{V}$ has the same units as ρ, which is necessary for Eq. (21).

3.2. *Comparison between Friedmann's Second Equation and Newton's Law of Motion*

Newton's Law of Motion for acceleration due to gravity force, using the force of gravity on mass m a distance R from a mass M, $F_g = mMG/R^2 = m \times$ acceleraton,

$$\ddot{R}(t) = \text{acceleration of force of gravity towards } M = \frac{MG}{R^2}, \tag{23}$$

with M = mass = density × volume = $\rho \times \frac{4\pi}{3}R^3$.

Therefore, with the direction towards M being inward (negative)

$$\ddot{R}(t) = -\frac{4\pi}{3}\frac{R^3}{R^2}G\rho$$

$$= -\frac{4\pi}{3}G\rho R$$

$$\frac{\ddot{R}(t)}{R(t)} = -\frac{4\pi}{3}G\rho,$$

which is Friedman's equation without pressure. Since from Eq. (22) pressure = energy due to force/volume, $\rho + 3p$ in Friedman's equation includes mass energy + energy due to force.

3.3. *Hubble's Parameter, Critical Density, and Curvature of Space*

The Hubble's parameter $H(t)$ is defined as $H(t) \equiv \dot{R}(t)/R(t)$. From this the Friedmann's equation, Eq. (19), can be written as

$$\frac{k}{H^2 R^2} = \frac{\rho}{3H^2/8\pi G} - 1 \equiv \Omega - 1, \tag{24}$$

with Ω the ratio of the density ρ to the critical density ρ_c:

$$\Omega = \frac{\rho}{\rho_c}$$

$$\rho_c \equiv \frac{3H^2}{8\pi G}$$

$$k = 1 \rightarrow \Omega > 1 \text{ closed universe} \tag{25}$$

$$k = 0 \rightarrow \Omega = 1 \text{ flat universe}$$

$$k = -1 \rightarrow \Omega < 1 \text{ open universe.} \tag{26}$$

When we study the CMBR we shall learn the value of Ω.

3.4. *Solution for R(t) Using Thermodynamics*

We use the First Law of Thermodynamics, with p = pressure, and the equation of state relating pressure to density.

The First Law of Thermodynamics is:

$$\text{change in internal energy} = -p \times \text{change in volume.} \tag{27}$$

The equation of state for the universe can be written as

$$p = w\rho,$$

where w is a constant relating pressure to energy density which depends on the nature of the universe at a particular time.

Using Eq. (27), we now derive the relationship between $R(t)$ and ρ, and from this the dependence of temperature of the universe on time, $T(t) \propto t^{-1/2} R(t)$ is found for different times in the universe (w is different for early times vs later times).

Since the internal energy = the energy density times the volume = ρV, $V = (4\pi/3)R^3$, the First Law, Eq. (27) can be written as

$$d(\rho R^3) = -pdR^3 = -d(pR^3) + R^3 dp$$
$$= -wd(\rho R^3) + wR^3 d\rho \quad \text{or}$$
$$d[(1+w)\rho R^3] = wR^3 d\rho. \tag{28}$$

The solution to Eq. (28) is

$$\rho \propto R^{-3(1+w)} \quad \text{or}$$
$$\rho = cR^{-3(1+w)}, \tag{29}$$

with c a constant.

To prove that ρ given in Eq. (29) is a solution to the First Law given in Eq. (28), note that

$$d[(1+w)\rho R^3] = c(1+w)[R^2(-3(1+w))R^{-3(1+w)}$$
$$+ R^{-3(1+w)}3R^2]dR$$
$$= wR^3(3c(1+w)R^{-3(1+w)}/R)dR$$
$$= wR^3 d\rho, \tag{30}$$

which proves that $\rho = cR^{-3(1+w)}$ is a solution to Eq. (28).

Another way to write Eq. (29) is

$$R \propto \rho^{-1/3(1+w)}. \tag{31}$$

In order to find how the radius of the universe depends on time, we need to find the t-dependence of p, which is done by finding its temperature dependence. Using the Einstein relationship ($p =$ momentum, not pressure) $p^2 = E^2 - m^2c^4$, giving energy $E = E(p)$,

$$\rho = \int d^3p E(p) f(p)$$
$$f(p) = \frac{1}{e^{(E-\mu)/T} \pm 1}, \tag{32}$$

where $f(p)$ is the phase space for fermions or bosons, as derived from thermodynamics. Carrying out the integral, at a time $= t$ one finds [3] for the temperature $= T(t)$,

$$\rho \propto T(t)^4 \propto t^{-2}$$
$$T(t) \propto t^{-1/2}. \tag{33}$$

From Eqs. (31), (33) one obtains the final result for the radius of the universe as a function of time:

$$R \propto t^{\frac{2}{3(1+w)}}. \tag{34}$$

3.5. $R(t)$ *During the Radiation Dominated and Matter Dominated Times*

For time $t < 1500$ years after the Big Bang the universe was dominated by radiation, photons rather than fermions and bosons. For $t > 1500$ years matter dominated the universe.

3.6. *Radiation and Matter Dominated Universe*

The very early universe was dominated by electromagnetic radiation, and in a E-dominated system the radiation pressure vs radiation density (equation of state) is

$$\text{radiation pressure} = \text{radiation density}/3 \quad \text{or}$$
$$p = \rho/3 \quad \text{or}$$
$$w = \frac{1}{3}.$$

Thus using $\rho = \text{constant} \times R^{-3(1+w)}$, for the E (radiation) dominated universe

$$\rho \propto \frac{1}{R^4} \quad \text{and}$$
$$R(t) \propto t^{1/2} = \sqrt{t}. \tag{35}$$

In the matter dominated universe there is matter density but no pressure, $p = 0$, therefore $w = 0$, giving

$$\rho \propto \frac{1}{R^3}$$
$$R(t) \propto t^{2/3}. \tag{36}$$

This is illustrated in the next figure. The time $= t_{eq} = 1500$ years is the time when the universe went from being radiation dominated to matter dominated.

RADIATION–MATTER RADIUS, R(t)

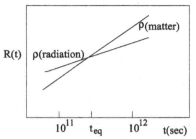

$t_{eq} \cong 1500$ years (radiation–matter equilibrium)

4. $T(t)$ from Friedmann's Equation

As was discussed above, Eq. (33), in the early radiation dominated universe the temperature dependence on time was $T(t) \propto 1/\sqrt{t}$. In Kolb and Turner [3] it was shown that

$$T(t) \simeq \frac{1\,\text{MeV}}{\sqrt{t(\text{in s})}}. \tag{37}$$

From this one finds results that are very important for the cosmology of the evolving universe:

$t = 10-100$ trillionth s, $T, E \simeq 300-100\,\text{Gev} \simeq$ mass of Higgs. EWPT, particles get mass.

$t = 10-100$ millionth s, $T, E \simeq 300-100\,\text{MeV} \simeq$ mass of pion $\simeq$ quark condensate, QCDPT, quarks condense to protons.

$t = 380{,}000$ years, $T, E \simeq 0.25\,\text{eV}$. Atoms form, universe has electric charge $\simeq 0$. Cosmic Microwave Background Radiation (CMBR) is released.

t greater than 1 billion years, stars, galaxies, and other structures in our present universe form.

$t = 13.8$ billion years, T is about 2.7 K.

5. The Horizon Problem and Inflation

The present size of our universe, given by the present Hubble's constant, $R_0 = H_0^{-1}$ is about 13.7 billion lightyears. From the well-known expansion rate, $\dot{R}$, one knows the size of the universe at an earlier time. Consider the very early time $t_i = 10^{-34}$ s. At that time the radius of our universe was

$$R(t_i) = L_i, \tag{38}$$

where the length L_i is a known from Friedmann's equation.

Since information cannot travel faster than the speed of light $= c$, the causally connected distance of our universe at this early time is

$$l_i = ct_i, \tag{39}$$

and at $t_i = 10^{-34}$ s,

$$L_i \gg l_i. \tag{40}$$

This means that at early times the matter making up our present universe was not causally connected. This is the HORIZON PROBLEM. How can we have a homogeneous, isotropic universe, which we will see ours is when we discuss the Cosmic Microwave Background Radiation (CMBR), if it is made from a huge number of disconnected mini universes?

SOLUTION: INFLATION

Recall the Friedmann's equation with no cosmological constant

$$\frac{\ddot{R}(t)}{R(t)} = -\frac{4\pi G}{3}(\rho + 3p)$$

$$\rho = \text{energy density} = \text{energy/volume}$$

$$p = \text{pressure}, \tag{41}$$

and that $\ddot{R}(t)$ is the acceleration of the rate of expansion, given by Hubble's constant.

Since both ρ and p are positive, Friedmann's equation in the form of Eq. (41) gives $\ddot{R}(t) < 0$. That is the expansion is slowing down due to the pull of gravity. Therefore there is no possible solution to the Horizon Problem with Eq. (41).

5.1. *Inflation via the Cosmological Constant,* Λ

Recall that Einstein introduced the cosmological constant about one hundred years ago to get a static universe, and then rejected it when Hubble showed that the universe is expanding.

A model that introduced the concept of an inflationary universe [4], Friedmann's equation with a comological constant Λ, is

$$\frac{\ddot{R}(t)}{R(t)} = -\frac{4\pi G}{3}(\rho + 3p) + \frac{\Lambda}{3}. \tag{42}$$

By choosing the parameters

$$p = -\rho = -\frac{\Lambda}{8\pi G}, \tag{43}$$

Eq. (42) becomes

$$\ddot{R}(t) = \frac{2\Lambda}{3}R(t). \tag{44}$$

Equation (44) is that of a classical harmonic oscillator, but with the opposite sign of the expected $\ddot{r}(t) = -kr(t)$, which would describe a vibrating spring, with a spring constant k. Taking two time derivatives one can easily prove that

$$R(t) = R(t = 0)e^{\sqrt{\frac{2\Lambda}{3}}t} \tag{45}$$

is the solution to Eq. (44).

With the current model of inflation and the value of Λ there is an exponential expansion of more than 10^{26} in 10^{-34} sec. A universe with a radius of 1 m would have a radius of 10^{26} m after inflation. In this short time the expansion was so enormous that the region which evolved and expanded for 14 billion years to form our current universe was causally connected, and almost completely homogeneous and isotropic. Inflation via the cosmological constant solves the horizon problem.

5.2. *Inflation via the Quintessence Field*

An alternate way of writing Eq. (43) is

$$\rho + 3p = -2\rho_{\text{VAC}}, \tag{46}$$

where ρ_{VAC} is the energy density of the vacuum. That is, if all matter were removed from the universe, and there was only a vacuum, there would still be an energy density $= \rho_{\text{VAC}}$.

Recent models of inflation [5, 6] introduce a new scalar field ϕ, the quintessence field, which is vacuum energy. The model introduced a potential $V(\phi)$, with ϕ satisfying the equation of motion

$$\ddot{\phi} + 3H\dot{\phi} + \frac{dV(\phi)}{d\phi} = 0. \tag{47}$$

The potential $V(\phi)$ is chosen to give inflation and other aspects of the universe, described below, in the 10^{-34} to 10^{-32} sec time interval.

Since friction, which causes a loss of energy and heating, is proportional to velocity $= \dot{r}$, the $H\dot{\phi}$ term corresponds to both the expansion of the universe, given by the Hubble's parameter H, and energy loss given by $\dot{\phi}$ that heats the universe.

For the current slow-roll model the potential $V(\phi)$ is shown in the figure below:

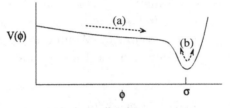

(a) is slow–roll region

(b) is region of oscillation about minimum of V

A crucial requirement of $V(\phi)$ is that the time for ϕ to increase from $\phi = 0$ at time $= 10^{-34}$ sec to its value $\phi = \sigma$ is very long compared to the expansion time. I.e.,

$$H\Delta t \gg 1. \tag{48}$$

This is accomplished by the slow-roll, with an almost constant potential in the region from $\phi = 0$ until just before it reaches the value of $\phi = \sigma$, shown as region (a) in the figure. During this period ϕ begins to dominate the energy density of the universe. It is Dark Energy, with a negative pressure causing inflation. Therefore the radius inflates

$$R(t) \propto e^{Ht}$$
$$H^2 = \frac{8\pi G}{3} V(\phi = 0), \tag{49}$$

with a result similar to that of the cosmological constant, Eq. (45). Then the potential decreases and increases, giving a minimum, so the quintessence field oscillates in the region of the minimum, region (b) in the figure. Due to the $\dot{\phi}$, the magnitude of ϕ quickly decreases to zero.

Therefore, the vacuum energy becomes equal to zero at the time $t = 10^{-32}$ sec. A problem with this scenerio is that with analysis of the CMBR, which we discuss in the next chapter, and recent measurements discussed in later chapters, vacuum energy (Dark Energy) is the dominant energy in our present universe. Thus we are far from understanding the nature of dark energy.

6. Gravitational Radiation and Gravity Waves from Inflation

Recall the discussion of the metric tensor $g_{\mu\nu}(x)$ for General Relativity compared to that for flat space, $g_{\mu\nu}(g_{00} = -1, g_{ii} = 1$ for $i = 1, 2, 3, g_{\mu\nu} = 0$ if $\mu \neq \nu)$. Following Weinberg [7], we define the metric $h_{\mu\nu}$ as the difference between the Einstein metric and that of flat space,

$$g_{\mu\nu}(x) = g_{\mu\nu} + h_{\mu\nu}, \tag{50}$$

and in real space the equation for gravitational radiation is [7]

$$\left[\Delta^2 - \frac{\partial^2}{\partial t^2} \right] h_{ij}(x, t) = -16\pi G S_{ij}(x, t), \tag{51}$$

with

$$S_{ij} = T_{ij}(x, t) - \frac{1}{3} \delta_{ij} \sum_{k=1}^{3} T_k^k(x, t), \tag{52}$$

where T_{ij} are the spatial components of the energy-momentum tensor.

The gravitational waves produced by primordial turbulence, including early universe inflation, has been estimated [8]. The magnitude of the radiation is too small to be detected at the present time. In the later chapters, when we discuss cosmological phase transitions, the gravity waves from the EWPT and QCDPT will be given and discussed.

7. Gravitational Quantum Field Theory

As discussed in Chapter 2, gravitational field theory is not part of the Standard Model. The quantum of the gravitational field, the graviton, differs from the quanta of the electromagnetic, strong, and weak fields in that its quantum spin is 2, rather than 1. Therefore, the field theory equations and methods are different, and cannot be applied. On the other hand, possible models have been published [7].

One of the most interesting aspects is the possible gravitational radiation emitted from atoms. Recall the discussion in Chapter 2 of the Bohr model, which introduced the concept of quantum states to explain the spectra of electromagnetic waves emitted from atoms.

Consider the energy levels of the hydrogen atom, which are labelled by nl, with n the principal and l the angular momentum quantum number; and $l = 0, 1, 2$ called s, p, d, as in the figure.

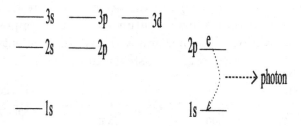

The electron in the 2p drops to the 1s state emitting a photon with frequency $\omega_{2p-1s} = 6.27 \times 10^8 \sec^{-1}$ or time to radiate 1.6×10^{-9} sec. The graviton radiation from the 3d to the 1s state is shown in the figure below:

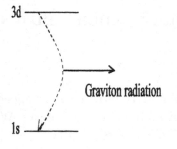

Since a graviton has spin=2 it cannot connect the electron in the 2p (l=1) to the 1s (l=0) state, but can connect the electron in the 3d (l=2) to the 1s state.

The graviton radiation emission rate (see Weinberg [7] for derivation) is $\Gamma_{3d-1s} = 2.5 \times 10^{-44}\,\text{sec}^{-1}$, which is so small that there is no possibility to observe gravitational radiation from atoms.

8. Einstein vs Bohr: Relativity vs Quantum Mechanics

Einstein, who introduced quantum concepts with his famous paper on the photoelectric effect, did not believe in quantum theory. Among his problems was Heisenberg's uncertainty principle, that one cannot measure both the position and momentum of an object. In the 1930s Einstein and Bohr had a famous discussion on the consistency of quantum theory and the Special Theory of Relativity. We briefly describe this.

CONCEPTS:

Electrons and positrons have quantum spin 1/2. A spin 0 particle (angular momentum = 0) at rest decays into an electron and positron, which have opposite momentum and spins in opposite direction from conservation of momentum and angular momentum. Each is detected 1 km from the origin, as shown in the figure.

EINSTEIN: Measure electron spin, find it is up. Instantly you know that the positron has spin down. Information travels 2 km in zero time, violating the theory of relativity. QM IS WRONG!

BOHR: It is true that the positron must have spin down, but there is no way to deliver that information over 2 km instantaneously. NO CONFLICT BETWEEN QM AND RELATIVITY.

VOTE BY PHYSICISTS: SORRY ALBERT, BUT NIELS BOHR IS CORRECT.

9. Problems

1a. If an object has mass = 1 kg and is moving with speed $v = 0.9c$, what is its momentum in nonrelativistic theory? What is its momentum in the Special Theory of Relativity?

1b. What is the relativistic energy (including mass energy as in Einstein's equation for E)?

2. What is the Schwarzschild radius for a mass 300 times the mass of the sun?

3a. Find the ratio of the radius of the universe, $R(t)$, for $t = 10^{11}$ s and $t = 10^{10}$ s, in a radiation dominated universe.

3b. Same problem as Problem 3a for $R(t = 10^{13}$ s$)/R(t = 10^{12}$ s$)$, in a matter dominated universe.

4. Using the approximate Kolb–Turner formula for $T(t)$, what is the temperature of the universe at time $t = 10^{-10}$ s? Compare this to the mass energy of the Higgs.

References

[1] Matts Roos, *Introduction to Cosmology*, John Wiley & Sons, Inc.

[2] Christian Møller, *The Theory of Relativity*, Clarendon Press, Oxford (1972).

[3] Edward W. Kolb and Michael S. Turner, *The Early Universe*, Addison-Wesley Publishing Co., (1990).

[4] Alan H. Guth, *Phys. Rev. D* **23**, 347 (1981).

[5] Ivaylo Zlatev, Limin Wang, and Paul J. Steinhardt, *Phys. Rev. Lett.* **82**, 896 (1999).

[6] Paul J. Steinhardt, Limin Wang, and Ivaylo Zlatev, *Phys. Rev. D* **59**, 123504 (1999).

[7] Steven Weinberg, *Gravitation and Cosmology: Principles and Applications of the General Theory of Relativity*, John Wiley & Sons, Inc. (1972).

[8] Tina Kahniashvili, Leonardo Campanelli, Grigol Gogoberidze, Yurii Maravi, and Bharat Ratra, *Phys. Rev. D* **D 78**, 123006 (2008).

Chapter 8

Cosmic Microwave Background Radiation (CMBR)

Below are some questions concerning the CMBR with answers that will be given in detail in this chapter.

QUESTION: What is Cosmic Microwave Background Radiation?

ANSWER: It is the electromagnetic radiation from the early universe, with a wavelength similar to that emitted from a microwave oven.

QUESTION: How does one measure the temperature of the universe?

ANSWER: Measure the amplitude of light in the sky as a function of wavelength.

QUESTION: How does one determine the amount and type of matter in the universe and test the ideas related to inflation?

ANSWER: Measure the light from the sky in all directions with two microwave telescopes and look for acoustic oscillations of photons at early times.

QUESTION: But do not the light from galaxies and other objects in the universe get in the way of the background radiation?

ANSWER: Yes, one must subtract this background.

1. Review of Concepts Relevant to the CMBR

In this section we briefly review topics related to cosmological parameters that are measured via the study of the Cosmic Microwave Background Radiation.

1.1. *Hubble's Parameter, $H(t)$*

Hubble and coworkers measured the velocity of a galaxy, v_g as a function of the distance, d_g of the galaxy from us using the Doppler shift, and found that velocity of a galaxy is proportional to d_g,

$$v_g = H(t)d_g,$$

where $H(t)$ is the Hubble's parameter, with $R(t) = c/H(t)$ the radius of the universe at time $= t$. $H_0 =$ the Hubble's constant at the present time has been measured by the CMBR, as we shall see.

1.2. *Centripetal Acceleration, Galaxy Rotation, and Dark Matter*

The centripetal acceleration, a_c, of a mass m rotating with constant speed $= v$ in a circle of radius R is

$$a_c = \frac{v^2}{R}.$$

If the centripetal acceleration of a mass m is given by the force of gravity F_g due to a mass M at a distance R

$$F(\text{gravity}) = G\frac{mM}{R^2}$$

then M must be large enough so that m has velocity v. Consider our Milky Way galaxy, shown in the next figure.

From the angular velocity and size of our galaxy, one can determine the mass in the galaxy. There is much more mass than visible mass, like stars. This is how Dark Matter was first discovered. From the CMBR we will learn how much matter is Dark Matter and how much is ordinary matter, like atoms, neutrinos...

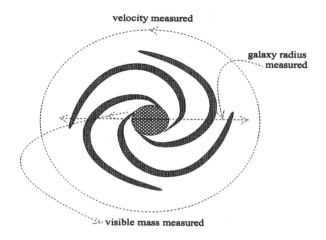

1.3. *Einstein's Theory of Gravity, Inflation, and Dark Energy*

From Einstein's equation of General Relativity one can determine the radius of the universe at time t, $R(t)$. From Einstein's Special Theory of Relativity (information cannot be sent faster than c = speed of light) one knows the largest distance which is causally connected (information can be sent). There is a problem at very early time, the Horizon Problem.

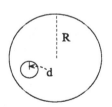

time = t_o=10^{-34} seconds

$R(t_0)$=radius of universe at time = t_o

d_0=ct_o =causally connected distance

$R(t_0) \gg d_0$ Problem. If our universe is made of many universes, it would not be homogeneous, as the CMBR finds.

The solution to the Horizon Problem is inflation. If one adds a cosmological constant Λ to Einstein's equation (Friedmann's version), or use a quintessence field model, the distance $d(t)$ is

$$d(t) = d_o e^{\sqrt{\frac{2\Lambda}{3}}(t-t_o)},$$

which results in a $d(t)$ much larger than $R(t)$ at $t = 10^{-32}$ seconds, so the early universe expands into uniform matter. This solves the problem that the CMBR finds our universe homogeneous, after taking into account galaxies.

Inflation theory predicts that the vacuum has energy, ρ_{VAC}, which is related to the cosmological constant Λ by (see Chapter 7, Section 5)

$$\rho_{VAC} = \frac{\Lambda}{8\pi G}. \tag{1}$$

As we shall see, the CMBR determines what fraction of the total energy density in the universe is Dark Energy = vacuum energy.

1.4. *k, the Curvature of Space*

The curvature constant k is given by

$$\frac{k}{H^2 R^2} = \Omega - 1, \tag{2}$$

with Ω the ratio of the density ρ to the critical density ρ_c:

$$\Omega = \frac{\rho}{\rho_c}. \tag{3}$$

As we shall soon see, the CMBR measures Ω and finds $\rho \simeq \rho_c$.

2. The CMBR

We now discuss certain concepts related to the Cosmic Microwave Background Radiation.

2.1. *Temperature and Time of CMBR Emission*

In the early universe the kinetic energy of electrons was too large for them to be bounded to atoms. Light at that time scattered from the free electrons. Light from the early universe was trapped. Due to the spread of energy of electrons at a given temperature, the heat energy, $kT(t)$ must drop to about $0.25\,\mathrm{eV}$ for electrons to be bounded.

The question is at what time did the universe become charge neutral, so light did not scatter and was free to be emitted.

The answer is that $t = 380,000$ years. To prove this, recall the formula for $T(t)$ derived from the Friedmann's version of Einstein's equation of General Relativity,

$$kT(t) \simeq \frac{1\,\text{MeV}}{\sqrt{t(\text{in s})}}, \tag{4}$$

where k is the Boltzmann's constant, and kT is energy. From this one can find kT at $t = 380,000$ years. Using 1 year $= 3.15 \times 10^7$ sec, as we saw in Chapter 1,

$$\sqrt{380,000 \text{ years in } s} = \sqrt{12.0 \times 10^{12}} = 3.46 \times 10^6. \tag{5}$$

From Eqs. (4), (5), using $1\,\text{MeV} = 10^6\,\text{eV}$,

$$kT(t = 380,000 \text{ years}) \simeq \frac{10^6\,\text{eV}}{3.46 \times 10^6} \simeq 0.25\,\text{eV}. \tag{6}$$

Therefore at $t = 380,000$ years with $kT \simeq 1/4\,\text{eV}$, electrons became bounded to the atomic nuclei and atoms were formed. The universe became electric charge neutral. Light from the early universe was released: the CMBR.

It is this CMBR, Cosmic Microwave Background Radiation, that we now discuss. As we shall see, we can learn very much about the early universe by an analysis (somewhat complicated) of the CMBR.

2.2. *How to Determine the Temperature of a Black Body*

A black body, like an oven or a microwave oven with a very small opening, emits electromagnetic radiation, like light, but with a much longer wavelength. We call it microwave radiation, as in CMBR. The figure below illustrates such a black body radiating.

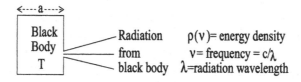

2.2.1. *Classical Rayleigh–Jeans Formula*

In classical physics the number of states for a single particle is given by phase space and n_s, the number of orientations of the spin of the particle. Therefore the number of state in dV with momentum between p and $p + dp$ is

$$N(p_i) = n_s \frac{dx\,dy\,dz\,dp_x\,dp_y\,dp_z}{h^3}, \tag{7}$$

or in a symmetric volume V, the spatial angle integration giving 4π, and using $n_s = 2$ as spin 1/2 electrons have two directions of spin, as we discussed in the previous chapter,

$$N(p) = \frac{8\pi V}{h^3} p^2 dp. \tag{8}$$

Using the relation $p = h/\lambda = h\nu/c$, with λ and ν the wavelength and frequency of light emitted from the cavity,

$$N(\nu) = \frac{8\pi V \nu^2 d\nu}{c^3}, \tag{9}$$

for a 3-dimensional cavity with volume V. Since for temperature T the average energy is $\bar{E} = kT$, as was discussed in Chapter 1, the energy as a function of frequency is

$$E(\nu) = N(\nu)kT = \frac{8\pi V \nu^2 d\nu}{c^3} kT. \tag{10}$$

Therefore the energy density for frequency ν, $\rho(\nu)$ is

$$\rho(\nu)_{R-J} = \frac{8\pi \nu^2}{c^3} kT, \tag{11}$$

which is the classical Rayleigh–Jeans formula. In the figure below $p(\nu)$ is shown for temperature $T = 1500K$.

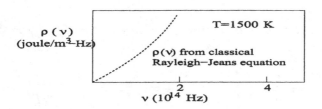

However, the classical Rayleigh–Jeans formula is not correct. Light travels like a wave, but has energy like a particle, a photon, with $E(\nu) = h\nu$ as we discussed previously.

2.2.2. *Quantum Theory: Rayleigh–Jeans ⇒ Planck Spectrum*

Planck realized that the classical theory was not correct for the energy associated with radiation. The relation for electromagnetic energy density used by Rayleigh–Jeans,

$$\rho(\nu)_{R-J} \propto \nu^2 kT \tag{12}$$

is not consistent with quantum theory. We now give a detailed derivation of the Planck spectrum, which is based on quantum theory.

The energy of light is given by the number of photons, n, times the energy of a photon for light with frequency ν, $E_n(\nu) = nh\nu$. Recall from the derivation of the Rayleigh–Jeans spectrum, the classical mean energy with temperature T is $\bar{E} = kT$, which is derived from the Boltzmann distribution,

$$P(E) = \frac{e^{-E/kT}}{kT} \tag{13}$$

where $P(E)dE$ is the probability of finding the system with energy between E and $E + dE$. Using the fact that in quantum theory the energy for frequency ν is $E_n(\nu) = nh\nu$,

$$P_n(E) = \frac{e^{-E_n/kT}}{kT}, \tag{14}$$

gives the energy probability for quantum systems. The mean energy $\bar{E}$ for temperature $= T$ is found by

$$\bar{E} = \frac{\sum_{n=0}^{\infty} E_n P_n(E)}{\sum_{n=0}^{\infty} P_n(E)} = \frac{\sum_{n=0}^{\infty} nh\nu e^{-nh\nu/kT}}{\sum_{n=0}^{\infty} e^{-nh\nu/kT}} = kT \frac{\sum_{n=0}^{\infty} nae^{-na}}{\sum_{n=0}^{\infty} e^{-na}}, \tag{15}$$

with $a = h\nu/kT$. Using the relationship $(d/dx)\ln(f(x)) = \frac{(d/dx)f(x)}{f(x)}$ and $(d/dx)(e^{-nx}) = -ne^{-nx}$

$$-a\frac{d}{da}\ln\left(\sum_{n=0}^{\infty} e^{-na}\right) = \frac{\sum_{n=0}^{\infty} nae^{-na}}{\sum_{n=0}^{\infty} e^{-na}} = \frac{\bar{E}}{kT}. \tag{16}$$

Next note that

$$\sum_n e^{-na} = 1 + e^{-a} + e^{-2a} + \cdots = 1 + e^{-a} + (e^{-a})^2 + \cdots$$

$$= \frac{1}{1 - e^{-a}}. \tag{17}$$

From Eqs. (15), (16), (17) one finds, using $kTa = h\nu$ and $e^a e^{-a} = 1$,

$$\bar{E} = -h\nu \frac{d}{da} \ln \left(\frac{1}{1 - e^{-a}} \right) = h\nu \frac{e^{-a}}{1 - e^{-a}} = \frac{h\nu}{e^{h\nu/kT} - 1}. \tag{18}$$

Therefore one finds for the Planck spectrum for $\rho(\nu)$, with $\bar{E}$ from Eq. (18) replacing kT for the R–J spectrum, Eq. (11),

$$\rho^{\text{Planck}}(\nu) = \frac{8\pi\nu^2}{c^3} \frac{h\nu}{e^{h\nu/kT} - 1}. \tag{19}$$

Also important for determining the temperature T of the cavity is a formula derived in 1893 by Wihelm Wien, who used the relationship between heat and electromagnetism to find the T from the maximum in $\rho(\nu)$, for frequency ν_{max}, which corresponds to the wavelength $\lambda_{\text{max}} = c/\nu_{\text{max}}$. This results in Wien's Law. The Planck spectrum and Wien's Law are summarized below.

PLANCK'S BLACKBODY SPECTRUM

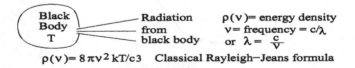

$\rho(\text{v}) = 8\pi\text{v}^2 kT/c3$ Classical Rayleigh–Jeans formula

Quantum: Photon Energy $= h\nu$.
Planck's spectrum for $\rho(\nu)$:

$$\rho(\nu) = \frac{8\pi^2\nu^2}{c^3} \frac{h\nu}{e^{h\nu/kT} - 1}$$

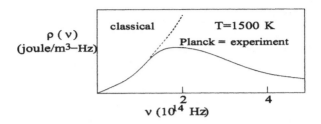

Wien's Law for wavelength at maximum:

$$\lambda_{\max} = \frac{0.29\,\text{cm}}{T(\text{in K})}$$

Therefore, by measuring ν, λ of the spectrum, one can determine T. Note if $h\nu \ll kT$, Planck spectrum = Rayleigh–Jeans spectrum.

Planck's blackbody spectrum is used for determining the temperature and temperature correlations from the CMBR, which we now discuss.

Using the Planck spectrum Penzias and Wilson with a radio telescope were the first to (approximately) determine the temperature of the universe. The figure is from their 1965 publication.

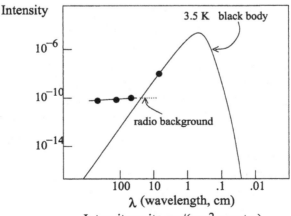

2.3. CMBR Temperature Measurements

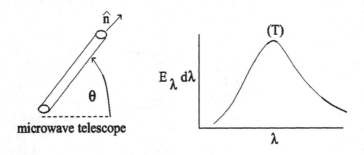

Observe T = temperature map using blackbody spectrum amplitude $A^T(\nu)$ (Planck spectrum with $\nu = c/\lambda$), using a microwave telescope in the $\hat{n}$ direction:

$$A^T(\nu, \hat{n}) = \frac{2hc^2\nu^3}{e^{h\nu/kT(\hat{n})} - 1} \tag{20}$$

Expand $T(\hat{n}) = T$ in $\hat{n}$ direction in Legendre polynomials, with $\hat{n}$ at an angle of θ

$$\frac{T(\hat{n})}{T_o} = 1 + \sum_l \sqrt{\frac{2l+1}{4\pi}} a_l^T P_l(\cos\theta), \tag{21}$$

with $P_l(\cos\theta)$ = Legendre polynomials, see the next subsection. T_o is T today, 14 billion years after the Big Bang.

Note that
(1) $\Delta T \equiv T - T_o$ = temperature fluctuations given by a_l^T
(2) T-fluctuations are random: $\langle a_l^T \rangle = 0$, where $\langle \ \rangle$ means take an average
(3) The main result of the measurement of $T(\hat{n})$ is finding T_o, the temperature of the universe at the present time:

$$T_o = 2.73 \, \text{K}, \tag{22}$$

which is an important result from CMBR.

2.4. *CMBR Temperature Correlations*

CMBR temperature correlations use two microwave telescopes in the $\hat{n}_1$ and $\hat{n}_2$ directions:

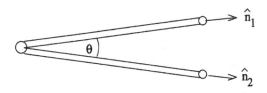

Expand the average of temperature differences, $\Delta T(\hat{n}_1) = T(\hat{n}_1) - T_o$ times $\Delta T(\hat{n}_2)$ in terms of $P_l(\cos\theta)$, where θ is the angle between $\hat{n}_1$ and $\hat{n}_2$:

$$\left\langle \frac{\Delta T(\hat{n}_1)}{T_0} \frac{\Delta T(\hat{n}_2)}{T_0} \right\rangle$$

$$= \sum_l C_l^{TT} \frac{2l+1}{4\pi} P_l(\cos\theta) \quad \text{where } \langle a_{l'}^T a_l^T \rangle = C_l^{TT} \delta_{ll'}.$$

The P_l are well-known Legendre polynomials:

$$P_0(z) = 1$$
$$P_1(z) = z$$
$$P_2(z) = 1 - 3z^2$$
$$P_3(z) = 3z - 5z^3,$$
$$\text{etc.}$$

The C_l^{TT} are the measured temperature–temperature correlations for the various values of the partial wave numbers l. From these one learns a great deal about the universe. See Roos [1] for a discussion of measuring T and the C_l^{TT} from the CMBR.

2.5. *CMBR TE, TB, EE, and BB Correlations*

In addition to the TT correlations, the CMBR can measure correlations associated with the polarization of the background electromagnetic radiation. Since an electromagnetic wave is transverse, (see first figure in Chapter 3), if the wave has frequency ω and is travelling in the z-direction, it has only x- and y-components (see first figure in Chapter 3).

$$E_x(t) = a_x(t)\cos(\omega t)$$
$$E_y(t) = a_y(t)\cos(\omega t). \tag{23}$$

If these two components are correlated, the electromagnetic wave is polarized. It is convenient to introduce the Stokes parameters, Q and U.

$$Q = a_x^2 - a_y^2$$
$$U = 2a_x a_y. \tag{24}$$

Similar to the a_l^T parameters defined for the expansion of $T(\hat{n})$ in Legendre polynomials, $P_l(\cos\theta)$, one can expand $(Q \pm iU)(\hat{n})$ as well as $T(\hat{n})$ in spherical harmonics, $Y_{lm}(x,y) \propto (d^m/dx^m)P_l(x)e^{imy}$, to obtain parameters $a_{lm}^T, a_{lm}^E, a_{lm}^B$:

$$T(\hat{n}) = \sum_{l,m} a_{l,m}^T Y_{lm}(\hat{n})$$

$$(Q + iU)(\hat{n}) = \sum_{l,m} a_{lm}^2 Y_{lm}(\hat{n})$$

$$(Q - iU)(\hat{n}) = \sum_{l,m} a_{lm}^3 Y_{lm}(\hat{n}) \tag{25}$$

$$a_{lm}^E = -(a_{lm}^2 + a_{lm}^3)/2$$
$$a_{lm}^B = i(a_{lm}^2 - a_{lm}^3)/2$$

The notation is E for electric and B for magnetic field.

Averaging over the measurements with two microwave telescopes and one obtains

$$\langle a_{l'm'}^T a_{lm}^T \rangle = C_l^{TT} \delta_{l'l}\delta_{m'm}$$
$$\langle a_{l'm'}^E a_{lm}^E \rangle = C_l^{EE} \delta_{l'l}\delta_{m'm}$$
$$\langle a_{l'm'}^B a_{lm}^B \rangle = C_l^{BB} \delta_{l'l}\delta_{m'm}$$
$$\langle a_{l'm'}^T a_{lm}^E \rangle = C_l^{TE} \delta_{l'l}\delta_{m'm}, \tag{26}$$

which have been measured by CMBR. Note BE and BT correlations $= 0$.

2.6. *Mass on a Spring: Harmonic Oscillator*

The source of temperature correlations, defined above, are oscillations of the dense fluid in the early universe, mainly a photon fluid as electromagnetic radiation dominated the early universe, which was used to find $R(t)$ before 1500 years. This oscillation is similar to the oscillations of a mass on a spring, a harmonic oscillator.

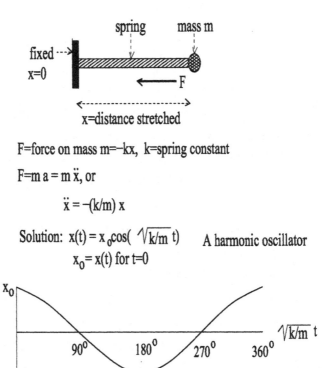

F=force on mass m=−kx, k=spring constant

F=m a = m ẍ, or

$$\ddot{x} = -(k/m)\, x$$

Solution: $x(t) = x_0 \cos(\sqrt{k/m}\ t)$ A harmonic oscillator

$x_0 = x(t)$ for t=0

2.7. *CMBR and Acoustic Oscillations*

Before recombination at 380,000 years the universe was a photon-baryon (neutrons and protons) fluid. Let us apply classical equations of fluid dynamics to find OSCILLATIONS OF THE FLUID. For a photon fluid it is convenient to work in momentum space, with momentum k related to the wavelength, λ.

TEMPERATURE AS A HARMONIC OSCILLATOR: For a pure photon fluid, the classical laws of fluid dynamics (Euler's equations, see Ref. [2]) show that the photon temperature, $T(t)$, for a given k, satisfies the differential equation (where c_s is a constant)

$$\ddot{T}_k(t) + c_s^2 k^2 T_k(t) = 0,$$

The solution to this equation is a classical harmonic oscillator:

$$T_k(t) = T_k(0) \cos(c_s k t)$$

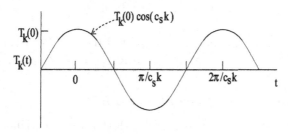

These acoustic oscillations are the mechanism for creating the C_l^{TT}, temperature–temperature correlations measured by the CMBR.

2.8. *Horizon Problem, Inflation and Acoustic Oscillations*

The horizon problem and the solution using inflation are briefly reviewed. Then how CMBR can test inflation is briefly discussed.

CMBR released at time t_d shows that two opposite regions of the sky are essentially identical.

Two regions in causal contact at the time $= t_o = 10^{-34}$ sec after the Big Bang would have lost all contact at time t_d, due to Einstein's axiom that information cannot travel faster than the speed of light.

HORIZON PROBLEM: How can regions initially not in thermal contact now be at the same temperature?

One solution was found using Friedmann's equation with a cosmological constant, Λ, which we found to be

$$\ddot{R}(t) = (2\Lambda/3)R(t) \text{ with the solution}$$

$$R(t) = R(t = 0)e^{\sqrt{\frac{2\Lambda}{3}}t}. \tag{27}$$

With the current value of Λ there is an expansion of more than 10^{26} in 10^{-32} seconds, solving the Horizon Problem. The constant Λ, provides a model for Dark Energy, the anti-gravity vacuum energy, producing inflation.

PREDICTION OF INFLATION: Due to this rapid expansion there will be acoustic oscillations of the photon fluid. The theoretical predictions using the quintessence field model discussed in the previous chapter, is that these acoustic oscillations will produce detectable temperature correlations, resulting in values of C_l^{TT} for large $l > 200$.

As we shall see, this has been detected by the CMBR, from which the fraction of Dark Energy in the universe has been found.

2.9. T_0 *Find* C_l^{TT} *at Recombination Time*

Starting from events (producing seeds) at an earlier time t, use the solutions to the Boltzmann equation:

$$C_l^{TT}(t_0) = \sum_{\text{wavelengths}} \sum_{t \to t_0} \text{scattering function } (t \to t_0)$$

$$\times \text{ source (at time} = t).$$

That is the Boltzman equation taking "seeds" from the time of some cosmological event to the time of last scattering (we call t_0) to obtain C_l^{TT} to compare to astrophysical observations. Some of the most important ones are:

COBE (COsmic Background Explorer): D.J. Fixsen *et al.*, *Astrophys. J.* **473**, 576 (1996).

WMAP (Wilkinson Microwave Anisotropy Probe): C.L. Bennett *et al.*, *Astrophysics Journal Supplement* **148**, 1 (2003), first year; N. Jarosik *et al.*, arxiv:[astro-ph] 1001.4744 (2010, seventh year); submitted to Astrophysics Journal Supplement.

ACBAR (Arcminute Cosmology Bolometer Array Receiver): C.L. Reichardt *et al.*, *Astrophys. J.* **674**, 1200 (2008).

QUaD (QUEST and DASI): S. Gupta *et al.*, arxiv: [astro-ph] 0909.1621 (2010). Measure EE and BB correlations as well as TT and TE correlations.

First Acoustic Peak of CMBR

The first acoustic peak, when $T(t)$ has its first maximum when expanded in the P_l angular functions, corresponds to $\cos(c_s k t)$ having a maximum. For a baryon-free fluid $c_s = c/\sqrt{3}$.

With units $c = 1$, this means that $kt/\sqrt{3} = \pi$ for the cosine to have a maximum. When expanded in P_l functions, one can show that a peak in the expansion coefficients, C_l^{TT} occurs — if the universe is flat, so $\Omega = 1$ — for $l \simeq 200$, as shown in the figure.

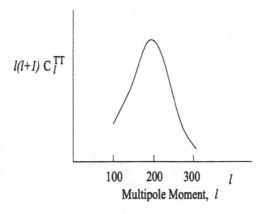

Apparatus used to measure CMBR correlations

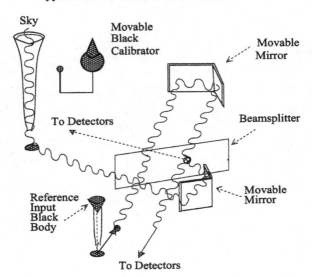

−200μK   200μK

The linearly cleaned WMAP team map (top), our Wiener filtered map (middle) and our raw map (bottom).

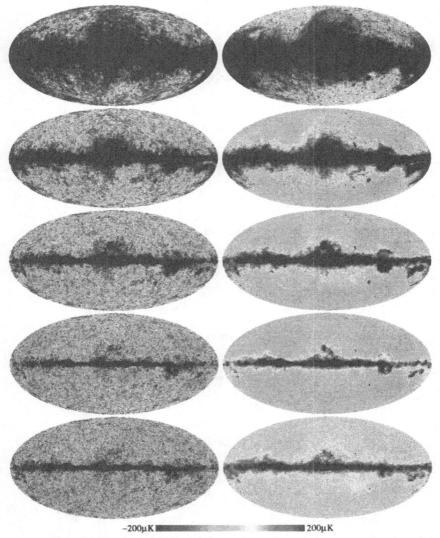

−200μK ▮▮▮▮▮▮▮▮▮▮▮▮▮▮▮▮▮▮▮ 200μK

The five WMAP frequency bands K, Ka, Q, V and W (top to bottom) before (left) and after (right) removing the CMB.

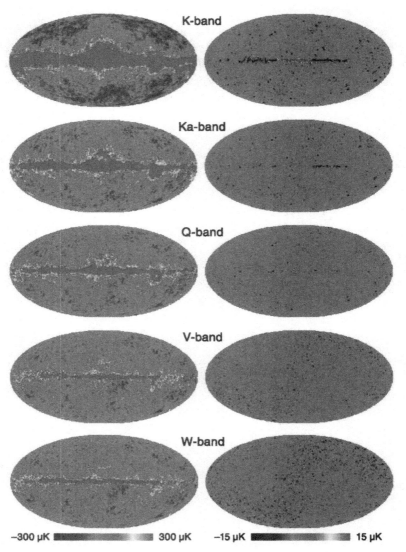

Plots of the Stokes I maps in Galactic coordinates. The left column displays the seven-year average maps, all of which have a common dipole signal removed. The right column displays the difference between the seven-year average maps and the previously published five-year average maps, adjusted to take into account the slightly different dipoles subtracted in the seven-year and five-year analyses and the slightly differing calibrations. All maps have been smoothed with a 1° FWHM Gaussian kernel. The small Galactic plane signal in the difference maps arises from the difference in calibration (0.1%) and beam symmetrization between the five-year and seven-year processing. Note that the temperature scale has been expanded by a factor of 20 for the difference maps.

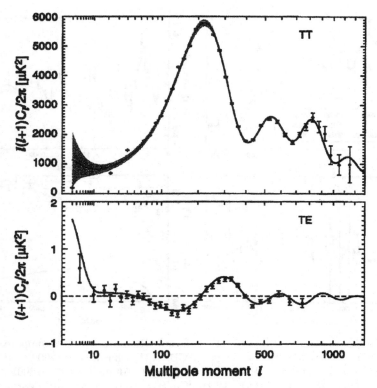

The temperature (TT) and temperature-polarisation (TE) power spectra for the seven-year WMAP data set. The solid lines show the predicted spectrum for the best-fit flat ΛCDM model. The error bars on the data points represent measurement errors while the shaded region indicates the uncertainty in the model spectrum arising from cosmic variance. The model parameters are: $\Omega_b h^2 = 0.02260 \pm 0.00053$, $\Omega_c h^2 = 0.1123 \pm 0.0035$, $\Omega_\Lambda = 0.728^{+0.015}_{-0.016}$, $n_s = 0.963 \pm 0.012$, $\tau = 0.087 \pm 0.014$ and $\sigma_s = 0.809 \pm 0.024$.

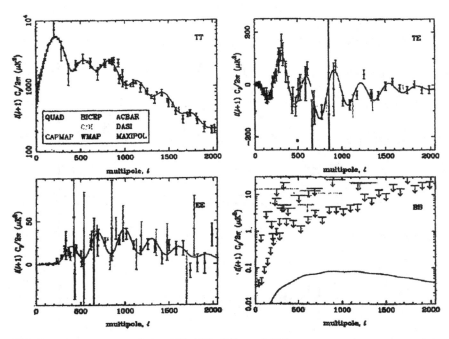

QUaD measurements of the TT, TE, EE, and BB power spectra compared to results from WMAP (Nolta *et al.*, 2009), ACBAR (Reichardt *et al.*, 2009), BICEP (Chiang *et al.*, 2009), B03 (Piscentitn *et al.*, 2006; Montroy *et al.*, 2006), CBI (Sievers *et al.*, 2007), CAPMAP (Bischoff *et al.*, 2008), MAXIPOL (Wu *et al.*, 2007), and DASI (Leitch *et al.*, 2005) experiments. The smooth curves in each panel are the power spectra expected in the best-fit ΛCDM model to the WMAP 5-year data.

Table 8.1.   *WMAP* Seven-year Cosmological Parameter Summary

Description	Symbol	WMAP-only	WMAP + BAO + H_0
Age of universe	t_0	13.75 ± 0.13 Gyr	13.75 ± 0.11 Gyr
Hubble constant	H_0	71.0 ± 2.5 km/s/Mpc	$70.4^{+1.3}_{-1.4}$ km/s/Mpc
Baryon density	Ω_b	0.0449 ± 0.0028	0.0456 ± 0.0016
Physical baryon density	$\Omega_b h^2$	$0.02258^{+0.00057}_{-0.00056}$	0.02260 ± 0.00053
Dark matter density	Ω_c	0.222 ± 0.026	0.227 ± 0.014
Physical dark matter density	$\Omega_c h^2$	0.1109 ± 0.0056	0.1123 ± 0.0035
Dark energy density	Ω_Λ	0.734 ± 0.029	$0.728^{+0.0015}_{-0.016}$

(*Continued*)

Table 8.1.   (*Continued*)

Description	Symbol	WMAP-only	WMAP + BAO + H_0
Curvature fluctuation amp- litude, $k_0 = 0.002 \, \text{Mpc}^{-1}$	$\Delta_{\mathcal{R}}^2$	$(2.43 \pm 0.11) \times 10^{-9}$	$(2.441^{+0.088}_{-0.092}) \times 10^{-9}$
Fluctuation amplitude at $8h^{-1} \, \text{Mpc}$	σ_8	0.801 ± 0.030	0.809 ± 0.024
Scalar spectral index	n_s	0.963 ± 0.014	0.963 ± 0.012
Redshift of matter–radiation equality	z_{eq}	3196^{+134}_{-133}	3232 ± 87
Angular diameter distance to matter–radiation eq.	$d_A(z_{eq})$	$14281^{+158}_{-161} \, \text{Mpc}$	$14238^{+128}_{-129} \, \text{Mpc}$
Redshift of decoupling	z_*	$1090.79^{+0.94}_{-0.92}$	$1090.89^{+0.68}_{-0.69}$
Age at decoupling	t_*	$379164^{+5187}_{-5243} \, \text{yr}$	$377730^{+3205}_{-3200} \, \text{yr}$
Angular diameter distance to decoupling	$d_A(z_*)$	$14116^{+160}_{-163} \, \text{Mpc}$	$14073^{+129}_{-130} \, \text{Mpc}$
Sound horizon at decoupling	$r_s(z_*)$	$146.6^{+1.5}_{-1.6} \, \text{Mpc}$	$146.2 \pm 1.1 \, \text{Mpc}$
Acoustic scale at decoupling	$l_A(z_*)$	302.44 ± 0.80	302.40 ± 0.73
Reionization optical depth	τ	0.088 ± 0.015	0.087 ± 0.014
Redshift of reionization	z_{reion}	10.5 ± 1.2	10.4 ± 1.2
Parameters for Extended Models[c]			
Total density	Ω_{tot}	$1.080^{+0.093}_{-0.071}$	$1.0023^{+0.0056}_{-0.0054}$
Equation of state	w	$-1.12^{+0.42}_{-0.43}$	-0.980 ± 0.053
Tensor to scalar ratio, $k_0 = 0.002 \, \text{Mpc}^{-1}$	r	$<0.36 \; (95\% \; \text{CL})$	$<0.24 \; (95\% \; \text{CL})$
Running of spectral index, $k_0 = 0.002 \, \text{Mpc}^{-1}$	$dn_s/d\ln k$	-0.034 ± 0.026	-0.022 ± 0.020
Neutrino density	$\Omega_\nu h^2$	$<0.014 \; (95\% \; \text{CL})$	$<0.0062 \; (95\% \; \text{CL})$
Neutrino mass	$\sum m_\nu$	$<1.3 \, \text{eV} \; (95\% \; \text{CL})$	$<0.58 \, \text{eV} \; (95\% \; \text{CL})$
Number of light neutrino families	N_{eff}	$>2.7 \; (95\% \; \text{CL})$	$4.34^{+0.86}_{-0.88}$

Results From CMBR Observations

$$\Omega = 1.0023^{+0.0056}_{-0.0054} \ (\text{Critical } \Omega_c = 1.0 \text{ for flat universe.})$$

Acoustic oscillations consistent with inflation

Dark energy density = 0.73 (vacuum energy)

Baryon density = 0.04 (4% of density)

Dark matter density = 0.23 (What is it?)

Number of neutrino flavors = $4.34^{+0.86}_{-0.88}$

Hubble constant = 71 km/s/Mpc

Age of universe = 13.7 billion years

3. Problems

1. Starting from the definition of $\bar{E}$, Eq. (15), carry out the derivation of $\bar{E}$, Eq. (18).
2. Prove that for $h\nu \ll kT$ the Planck spectrum for $\rho(\nu)$ is the same as the Rayleigh–Jeans spectrum.

3a. Prove that $x(t) = x_o \cos(bt)$ is a solution to the harmonic oscillator equation $\ddot{x}(t) = -b^2 x(t)$.

3b. Prove that $x(t) = x_o e^{bt}$ is a solution to the harmonic oscillator equation $\ddot{x}(t) = +b^2 x(t)$.

4. Using Eq. (21), and $a_0^T = 1$, $a_1^T = 0.25$, $a_2^T = 0.1$, plot $T(\hat{n})/T_0$ for $\cos(\theta) = 1$ to 0. I.e., $\theta = 0$ to 90 degrees.

References

[1] Matts Roos, *Introduction to Cosmology*, John Wiley & Sons, Inc.
[2] Steven Weinberg, *Gravitation and Cosmology: Principles and Applications of the General Theory of Relativity*, John Wiley & Sons, Inc. (1972).

Chapter 9

Electroweak Phase Transition (EWPT)

During a cosmological phase transition the vacuum of the universe changes to a different vacuum. There were two cosmological phase transitions, the Electroweak Phase Transition (EWPT) and the Quantum Chromodynamics Phase Transition (QCDPT).

In the present chapter we discuss the EWPT.

THE EVOLUTION OF THE UNIVERSE — COSMOLOGICAL PHASE TRANSITIONS

t = Time	T = Temperature	Events
10^{-35} s	10^{14} GeV	Big Bang, Strings, Inflation. Very early. Current particle theory no good.
EWPT → 10^{-11} s	100 GeV	Electroweak phase transition. Particles (Higgs) get masses. Particle Theory OK. Baryogenesis (more particles than antiparticles) needs one supersymmetric particle.
QCDPT → 10^{-5} s	100 MeV	QCD phase transition. Quantum Chromodynamics theory. Quark (Plasma) condenses to hadrons (protons...). OUR PRESENT UNIVERSE BEGINS.

1. Phase Transitions

A phase transition is the transformation of a system in thermodynamic equilibrium, with a well defined temperature for the entire system, from one phase of matter or state to another. We now discuss the two basic types of phase transitions: classical, when one phase transforms to a different phase; and quantum, when one state transforms to a different state.

1.1. *Classical Phase Transitions*

A classical phase transition is the transformation of a phase of a system to another phase. The three most common phases are solid, liquid, and gaseous; and also under special conditions there is a plasma phase.

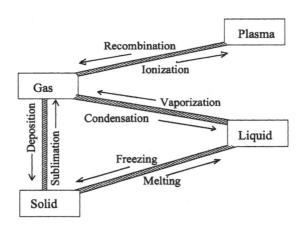

For applications to cosmology we are mainly interested in FIRST ORDER phase transitions. These phase transitions occur at a critical temperature, T_c, and the temperature stays constant until all the matter in the system changes into the new phase. For example, if one heats water (a liquid) at standard atmospheric pressure, it starts to boil, with bubbles of steam (a gas), and the temperature stays at 100°C.

The heat energy that turns the water to steam is called the latent heat.

First order vs higher order phase transitions

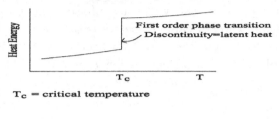

T_C = critical temperature

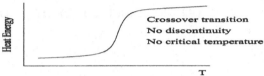

A familiar example of first order phase transitions is ice, a solid, melting to form water, a liquid, and water boiling to form steam, a gas. The figure below shows these two first order phase transitions for one gallon of water.

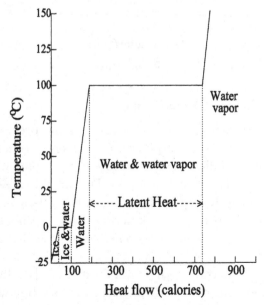

Phases of water with heat flowing into one gallon of water.

In the figure below is the phase diagram for water–ice–steam. At atmospheric pressure the water–ice phase transition occurs at $T = 0°$C and the water–steam phase transition at $T = 100°$C. For higher pressure the water–ice transition occurs for $T < 0°$C and the water–steam phase transition occurs for $T > 100°$C. At a very low pressure there is a triple point, and for a lower pressure there is no liquid phase.

PHASES OF WATER (LIQUID), ICE (SOLID), STEAM (GAS)

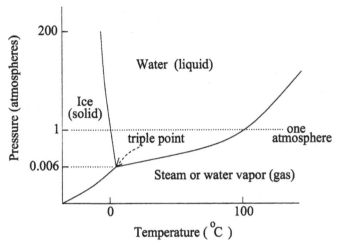

Phase diagram of water for pressure vs temperature.

1.2. *Quantum Phase Transitions*

A classical phase transition, as was treated in the previous subsection, is the transition from one phase of a system to another. By a phase, for example a liquid phase, the system has atoms which interact to form matter that we call a liquid. During the classical phase transition to a gas the heat added to the liquid causes the atoms to move faster, as the heat energy is kinetic energy of the atoms. This causes the atoms to separate and become a gas, with the heat energy called latent heat.

A quantum phase transition is very different [1]. In quantum mechanics one does not deal with physical matter, but with states. A quantum phase transition is the transition from one state to a different state. We now review some basic aspects of quantum mechanics needed for the study of cosmological phase transitions, such as the EWPT, an extension of the discussion in Chapter 5.

1.2.1. *Basic Aspects of Quantum Mechanics: States, Operators, Expectation Values*

In quantum theory one deals with states and operators. A quantum state represents the system, and a quantum operator operates on a state. For instance, a system is in state[1], and there is an operator A

$$|\text{state}[1]\rangle \equiv \text{state}[1]$$
$$A = \text{operator } A. \tag{1}$$

An operator operating on a quantum state produces another quantum state. For example operator A operates on state[1]:

$$A|\text{state}[1]\rangle = |\text{state}[2]\rangle, \tag{2}$$

where state[2] is another quantum state.

State[2] might also be the same as state[1]. Thus, calling $|\text{state}[1]\rangle = |A\rangle$,

$$A|A\rangle = a|A\rangle, \tag{3}$$

where a is called the eigenvalue of the operator A in the state $|A\rangle$. It is the exact value of A. If a state is not an eigenstate of an operator, the operator does not have an exact value. This is not a question of making a measurement, as in classical theory, but is a basic aspect of quantum theory.

For example, classically an electron has momentum $\vec{p}$. In quantum theory we would treat this simple system as a state $|e, \vec{p}\rangle$. Then the momentum operator $\vec{p}_{op}$ when operating on $|e, \vec{p}\rangle$

$$\vec{p}_{op}|e, \vec{p}\rangle = p|e, \vec{p}\rangle, \tag{4}$$

where $\vec{p}$ is the momentum of the electron. As we shall briefly discuss below, we know nothing of the position of the electron.

QUESTION: If the system is in state[1], what is the value of operator A?

$$\langle\text{state}[1]| \equiv \text{``adjoint'' of state}[1]$$

$$\langle\text{state}[1]|A|\text{state}[1]\rangle \equiv \text{EXPECTATION VALUE OF A}, \qquad (5)$$

where the expectation value is the average value of A that we would find if we carried out a measurement. Note that A does not have an exact value, unless state[1] is an eigenstate of A.

1.2.2. *Momentum and Position, Energy and Time, Hamiltonion*

As we discussed in the Chapter 1, the position vector is $\vec{r} = (x, y, z)$ and the momentum vector is $\vec{p} = (p_x, p_y, p_z)$ in classical physics. In quantum mechanics $\vec{r}$ and $\vec{p}$ are operators, with

$$p_x = \frac{\hbar}{i}\frac{d}{dx}, \qquad (6)$$

where $\hbar = h/2\pi$, with analogous relations for y and z. See Schiff [2] for a discussion of momentum and momentum eigenstates. Since

$$\frac{d}{dx}x = 1 + x\frac{d}{dx}$$

$$p_x x \neq x p_x, \qquad (7)$$

a state cannot be an eigenstate of both x and p_x. If one measures both x and p_x for some object, the uncertainty in $x = \Delta x$ and the uncertainty in $p_x = \Delta p_x$ satisfy the relationship

$$\Delta p_x \Delta x \geq \hbar/2, \qquad (8)$$

which is the Heisenberg Uncertainty Principle for position and momentum. The quantum mechanical energy operator is

$$E = i\hbar\frac{d}{dt}. \qquad (9)$$

Using a similar derivation as for position and momentum, this leads to the Heisenberg Uncertainty Principle for energy and time:

$$\Delta E \Delta t \geq \hbar. \qquad (10)$$

The operator for energy is the Hamiltonian $E = H$. Therefore,

$$H|E; \vec{r}, t\rangle = i\hbar \frac{d}{dt}|E; \vec{r}, t\rangle = E|E; \vec{r}, t\rangle, \tag{11}$$

where the energy eigenvalue E is the energy of the system.

Note that

$$\frac{d}{dt}e^{at} = ae^{at}. \tag{12}$$

It follows from Eqs. (11), (12) that the energy eigenstate $|E; \vec{r}, t\rangle$ has the form

$$|E; \vec{r}, t\rangle = e^{-iEt/\hbar}|E; \vec{r}\rangle, \tag{13}$$

which we used in the discussion of neutrino oscillations in Chapter 5.

1.3. Quantum Field Theory, Lagrangians, and Cosmological Phase Transitions

In Chapter 2 the Standard Model particles and forces were discussed. The force was shown via a Feynman diagram, for example, the electromagnetic force between electrons given by the exchange of a photon. This is not generally applicable. We now discuss method for treating quantum fields via the Lagrangian.

1.3.1. Lagrangian and Euler–Lagrange Equation

Quantum field theory starts with a Lagrangian, L, which is a function of the fields being considered [2]. For example, if one is considering only one scalar (quantum spin 0) field, like the Higgs field, ϕ, then $L = L(\phi, \partial_\mu \phi)$, where $\partial_\mu \phi$ is the derivative of ϕ with respect to x_μ. The Euler–Lagrange equation of motion is

$$\partial_\mu \frac{\partial L}{\partial(\partial_\mu \phi)} - \frac{\partial L}{\partial \phi} = 0. \tag{14}$$

As an example consider the Lagrangian $L(\phi, (d/dx)\phi), \phi = \phi(x)$,

$$L = \left(\frac{d\phi}{dx}\right)^2 - (\phi)^2 \tag{15}$$

$$\partial_\mu \frac{\partial L}{\partial(\partial_\mu \phi)} = 2\frac{d}{dx}\frac{d\phi}{dx} = 2\frac{d^2}{dx^2}\phi$$

$$\frac{\partial L}{\partial \Phi} = -2\phi. \tag{16}$$

From Eqs. (14), (15), (16) one obtains the equation of motion (EOM)

$$2\frac{d^2}{dx^2}\phi + 2\phi = 0. \tag{17}$$

Using $d\sin(x)/dx = \cos(x)$ and $d\cos(x)/dx = -\sin(x)$, Eq. (17) has the solution

$$\phi(x) = a \cdot \sin(x) + b \cdot \cos(x), \tag{18}$$

where a and b are constants determined by the initial conditions. For example if $\phi(0) = 0$ and $(d/dx)\phi(0) = 1$ then $b = 0$ and $a = 1$.

1.3.2. *Cosmological Phase Transitions*

Let us call $|0, T\rangle$ the state of the universe at some time when it has a temperature T. An operator A has the expectation value $= \langle 0, T|A|0, T\rangle$, as we discussed above. If there is a cosmological first order phase transition, then there is a critical temperature T_c, and $\langle 0, T|A|0, T\rangle_{T<T_c} - \langle 0, T|A|0, T\rangle_{T>T_c} = \Delta A =$ latent heat of the phase transition. As we shall discuss in this chapter, the operator A is the Higgs field, ϕ, for the EWPT, while it consists of quark, q, and antiquark, $\bar{q}$, fields for the QCDPT, as we discuss in the next chapter.

2. Electroweak Phase Transition (EWPT)

If the Electroweak Phase transition is first order:

(1) PARTICLES GET MASS,
(2) MAGNETIC FIELDS ARE CREATED,
(3) BARYOGENESIS — MORE QUARKS THAN ANTIQUARKS?

It has been shown, however, that with the Standard Model, if the mass of the Higgs is greater than about 60 GeV, the EWPT is not first order, it is a crossover. Since the Higgs mass is greater than 100 GeV, in the Standard Model not only is the EWPT a second order phase transition, but baryogenesis cannot occur. For this reason, a number of theorists have included a Supersymmetric particle in their theory.

2.1. *Review of Standard Electroweak Theory and Stop*

FERMIONS (spin 1/2 particles) are (e^-, ν_e) and the μ and τ leptons with their neutrino partners; and the quarks (q_u, q_d) and the other two quark generations.

GAUGE BOSONS (spin 1 particles) are the photon (electromagnetic field), gluon (strong chromodynamic field), and W^+, W^-, Z^0 (weak field).

HIGGS, ϕ, a scalar boson (spin 0) completes the Standard Model.

STOP, Φ_s: As was discussed above, the EWPT is not first order unless at least one boson in addition to those in the Standard Model is included. We discuss a Minimal Supersymmetric Model (MSSM), with the supersymmetric partner to the top quark, the stop added to the electroweak theory.

2.2. *First Order EWPT Using MSSM*

The EWPT is characterized by the vacuum value of the Higgs field = expectation value of ϕ in the vacuum state with temperature T, $|0, T\rangle$. If there is a first order EWPT at temperature $= T_c$

$$\langle 0, T | \phi | 0, T \rangle = 0 \quad \text{for } T > T_c$$
$$\langle 0, T | \phi | 0, T \rangle \neq 0 \quad \text{for } T < T_c.$$

For $T < T_c$, $\langle 0, T | \phi | 0, T \rangle \propto$ Higgs mass. Therefore, with a first order EWPT the Higgs mass goes from 0 to the nonzero M_H as the temperature drops through T_c, at about 10^{-11} s after the Big Bang.

A standard model for ϕ near $T \approx T_c$ is

$$\ddot{\phi} = \text{effects of the gauge fields} + V(\phi), \tag{19}$$

with the potential $V(\phi)$ changing with T as shown in the figure.

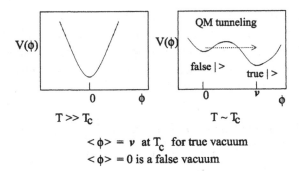

$$\langle\phi\rangle = v \text{ at } T_c \text{ for true vacuum}$$
$$\langle\phi\rangle = 0 \text{ is a false vacuum}$$

QM tunneling takes universe from false to true vacuum.

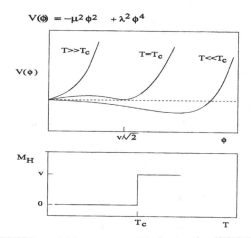

EWPT and Higgs mass with first order EWPT.

All particles except photon get mass, with M_H = mass of Higgs particle:

$$M_H = v$$
$$M_W = gv/\sqrt{2}, g = \text{strong coupling constant}$$
$$M_Z = M_W/\cos(\theta_W), \theta_W = \text{Weinberg angle}$$
$$m_e \propto m_u \propto m_d \propto v \qquad (20)$$

Standard Model: $M_W = 37\,\text{GeV}/\sin(\theta_W) \simeq 80\,\text{GeV}$

$$M_Z \simeq 90\,\text{GeV}$$

In 2012 experiments with the LHC found that $M_H \simeq 125\,\text{GeV}$.

2.3. *First Order EWPT: Bubbles Form*

During a first order cosmological phase transition, with $T = T_c$, bubbles with the new cosmological vacuum form within the vacuum for $T > T_c$.

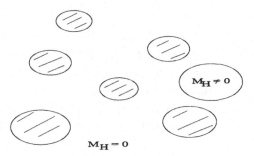

Bubbles of universe with $M_H \neq 0$ form in universe during the EWPT, with $M_H \neq 0$ and $\langle \phi \rangle \neq 0$ inside the bubbles, while $M = 0$ and $\langle \phi \rangle = 0$ outside the bubbles. That is, a new universe vacuum state is created by the EWPT.

Bubbles nucleate and then collide, creating electromagnetic fields and then magnetic fields.

During bubble nucleation electromagnetic fields form

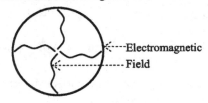

After bubbles collide and merge magnetic fields form

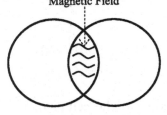

Seeds for galactic magnetic fields ?

2.4. *MSSM EW Lagrangian and Electromagnetic Field Creation During Nucleation*

The MSSM electroweak Lagrangian without fermions is

$$L^{MSSM} = L^{MSSM}(\phi, \Phi_s, W^i_\mu, A^{em}_\nu), \tag{21}$$

where $i = 1, 2, 3$, $W^1 = W^+, W^2 = W^-, W^3 = Z$ are the weak fields; and A^{em}_ν is the electromagnetic field, which were discussed earlier. L^{MSSM} is also a function of $\partial_\alpha \phi, \partial_\alpha \Phi_s, \partial^\nu A^{em}_\nu$ etc. This Lagrangian and the equations of motion are discussed in detail in Ref. [3]. We do not give the rather complicated details here, but discuss the final equations and the electromagnetic field created via EWPT bubble nucleation.

Those readers who do not wish to follow the derivation of the electromagnetic field can go to the results for the function $A_r(t)$ shown in Fig. 9.1 following the differential equations.

The potential in the Lagrangian for the Higgs is

$$V(\phi) = -\mu^2 \phi^2 + \lambda \phi^4. \tag{22}$$

with the parameters μ, λ chosen to give the EWPT and Higgs mass shown in the figure above. The potential for the Higgs-stop is

$$V_{hs}(\Phi) = -\mu_s^2 \Phi_s^2 + \lambda_s \Phi_s^4 + \lambda_{hs} \phi^2 \Phi_s^2, \tag{23}$$

with the parameters $\mu_s, \lambda_s, \lambda_{hs}$ chosen to give the EWPT and Higgs mass shown in the figure above.

As described in Ref. [3], since we are treating bubble nucleation with spherical symmetry a reasonable approximation for the W^i_ν, A^{em}_ν fields is

$$W^i_\nu = ix_\nu W(r, t)$$
$$A^{em}_\nu = ix_\nu A(r, t), \tag{24}$$

where r is the distance from the center of the bubble and $x^\mu x_\mu = t^2 - r^2$, with units $c = 1$. The W, A fields must also satisfy gauge conditions. See, e.g., Ref. [4] for a discussion of gauge conditions.

The gauge conditions for the W and A fields are

$$\sum_{j=1}^{3} \partial_j W^j = \sum_{j=1}^{3} \partial_j A^j = 0. \tag{25}$$

Using Eqs. (24), (25) the gauge conditions are

$$r\partial_r W(r,t) + 3W(r,t) = 0$$
$$r\partial_r A(r,t) + 3A(r,t) = 0, \tag{26}$$

with solutions for a given r = radius of bubble

$$W(r,t) = \frac{W_r(t)}{r^3}$$

$$A(r,t) = \frac{A_r(t)}{r^3}. \tag{27}$$

Using the MSSM EW Lagrangian L^{MSSM}, with the Euler–Lagrange equations (see Ref. [3]) one finds the differential equation for the functions $W_r(t)$, $A_r(t)$:

$$\ddot{W}_r r^2 - 3t\dot{W}_r + 3W_r + g\frac{t^2 - r^2}{r}W_r\left(\frac{\dot{W}_r}{t} - \frac{3W_r}{r^2}\right) = 0 \tag{28}$$

$$\ddot{A}_r r^2 - 3t\dot{A}_r + 3A_r + G\frac{t^2 - r^2}{r}W_r\left(\frac{\dot{W}_r}{t} - \frac{3W_r}{r^2}\right) = 0, \tag{29}$$

with $g = 0.646$ and $G = 0.303$.

One first finds solutions to Eq. (28) for $W_r(t)$ for a series of values of the bubble radius r, and use this in Eq. (29) to find $A_r(t)$. The results of Henley *et al.* [3] are shown in Fig. 9.1.

From the figure one observes that the time at which one reaches the radius of the bubble wall is given approximately by

$$t \simeq 2r, \tag{30}$$

from which we obtain the nucleation velocity of the bubble wall:

$$v^{wall} \simeq \frac{c}{2}. \tag{31}$$

This is an important result for the derivation of magnetic field generation during EW bubble collisions, discussed in the next subsection.

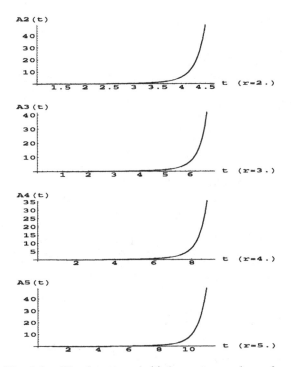

Fig. 9.1.   The function $A_r(t)$ for various values of r.

2.5. *Magnetic Field Creation During EWPT Bubble Collisions*

During nucleation, with spherical symmetry, the elecromagnetic field $A_\nu(r,t)$ is created, but with no direction in space. When the bubbles collide a direction is given, which we call z, as shown in the figure below.

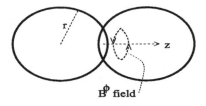

Because of the symmetry in the x- and y-directions it is useful to introduce a new coordinate τ, defined by

$$\tau^2 = t^2 - x^2 - y^2. \qquad (32)$$

Since without fermions the only fields with electric charge are the W^+, W^-, which we call W^a for convenience, one only needs $W_\nu^a(\tau, z)$. From symmetry $W_z^a(\tau, z)$ is

$$W_\nu^a(\tau, z) = x_\nu W^a(\tau, z). \tag{33}$$

From the Lagrangian the equations of motion are [5, 6]

$$\left(\frac{\partial^2}{\partial \tau^2} + \frac{2}{\tau}\frac{\partial}{\partial \tau} - \frac{\partial^2}{\partial z^2} + M^2\right) W_z^a = 0, \tag{34}$$

and

$$\left(\frac{\partial^2}{\partial \tau^2} + \frac{4}{\tau}\frac{\partial}{\partial \tau} - \frac{\partial^2}{\partial z^2} + M^2\right) W^a = 0. \tag{35}$$

The electromagnetic current $j_\nu = j_0$ ($\nu = 0$), $x_\nu j$ ($\nu = 1, 2$), j_z ($v = 3$) is given by W_z, W found by solving Eqs. (34), (35). See Ref. [5] for details.

Using $j(t, z)$ and $j_z(t, z)$ one obtains the magnetic field $\vec{B}$ from Maxwell's equations [4].

From symmetry, the B field is a B^ϕ field, in circles around the z-direction, as shown in the figure above. The equation used to find B^ϕ is

$$\left(\frac{\partial^2}{\partial t^2} - r\frac{\partial^2}{\partial r^2}\frac{1}{r} - 3\frac{\partial}{\partial r}\frac{1}{r} - \frac{\partial^2}{\partial z^2}\right) B^\phi = 4\pi \left(\frac{\partial j_z}{\partial r} + r\frac{\partial j}{\partial z}\right). \tag{36}$$

The solution for B^ϕ as a function of r for $z = 0$ for a series of times from Stevens *et al.* [5] is shown in the figure below.

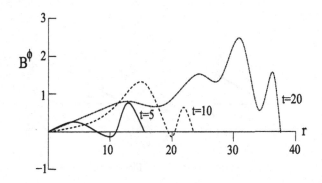

Finally, we need the B field at the end of the EWPT for calculation of gravitational radiation, which is discussed in Section 3.2.2. It is $B^{EWPT}(T_c) = 10M_W^2$, with $M_W \simeq 80\,\mathrm{GeV}$.

2.5.1. B^{EWPT} as Seed for Galaxy and Galaxy Cluster Magnetic Fields?

The origin of observed magnetic fields in galaxies and galaxy clusters is a mystery [7]. These fields from B^{EWPT} seeds do not fit the description [8], but seeds from the QCDPT might work [9]. We discuss this in the next chapter.

3. Gravitational Quantum Field Theory and Gravity Waves

As was pointed out in Chapter 2, although the force of gravity is our most familiar force, and the most important force in the theory of the evolution of the universe, Einstein's General Theory of Relativity is a classical field theory. Quantum field theory for gravity does not work. The reason for this is that the quantum of the gravitational field, the graviton, has spin 2, as we discuss below. First we discuss tensors compared to vectors, then the emission of gravitons, and finally gravity waves.

3.1. *Vectors and Tensors*

A vector has three components. For example, the position vector, $\vec{r}$, with components $r_i, i = 1, 2, 3$; or $r_1 = x$, $r_2 = y$, $r_3 = z$ and the electric field $\vec{E}$ has components E_x, E_y, E_z as shown in the figure below.

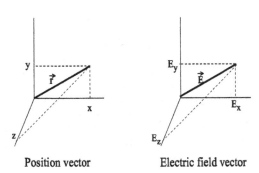

Position vector   Electric field vector

A nonrelativistic tensor has 3×3 components: T_{ij}, i or $j = 1, 2, 3$. A relativistic vector has four components, $V_\mu, \mu = 0, 1, 2, 3$, while a relativistic tensor has 4×4 components: $T_{\mu\nu}$ with μ and $\nu = 0$, $1, 2, 3$.

The vector fields are electromagnetic, strong, and weak. Quanta of vector fields are the photon, gluon, and (W^+, W^-, Z^0). All quanta of these vector fields have quantum spin $= 1$. As we now see, the gravitational field is a tensor field, and the quantum, the graviton, has quantum spin 2.

3.2. *Einstein's Gravitational Quantum Field Theory*

We first discuss Einstein's Theory of General Relativity, a classical field theory, and derive the Friedmann's equation used in our earlier derivation of the radius and temperature of the universe. We then discuss possible quantum field theory, and the graviton.

3.2.1. *Einstein's General Theory of Relativity*

In our earlier discussion of the General Theory of Relativity we used Friedmann's equations. We now discuss Einstein's original theory.

Einstein's tensor, $G_{\mu\nu}$ with $\mu, \nu = 0, 1, 2, 3$ is defined as [10]

$$G_{\mu\nu} = \mathcal{R}_{\mu\nu} - \mathcal{R}g_{\mu\nu}, \tag{37}$$

with $\mathcal{R}_{\mu\nu}$ and $g_{\mu\nu} = 0$ if $\mu \neq \nu$. For no space curvature, $k = 0$, $\mathcal{R}_{\mu\mu}$ is

$$\mathcal{R}_{ii} = \frac{\ddot{R}}{R} + 2\frac{(\dot{R})^2}{R^2}$$

$$\mathcal{R}_{00} = -3\frac{\ddot{R}}{R} \tag{38}$$

$$\mathcal{R} = -6\left[\frac{\ddot{R}}{R} + \frac{(\dot{R})^2}{R^2}\right],$$

where $i = 1, 2,$ or 3. The elements of the stress-energy tensor, $T_{\mu\nu}$, which are needed are given below.

Einstein's equation of General Relativity, without a cosmological constant, is:

$$G_{\mu\nu} = 8\pi G T_{\mu\nu}, \tag{39}$$

with $T_{\mu\nu}$ the stress-energy tensor.

Since our universe is spherically symmetric, one can show that the only nonvanishing elements of $T_{\mu\nu}$ are

$$T_{11} = T_{22} = T_{33} = p \text{ (pressure)}$$

$$T_{00} = \rho \text{ (energy density).} \tag{40}$$

From this and expressing $G_{\mu\nu}$ in terms of $R(t)$ one derives Friedmann's equation:

$$\frac{\ddot{R}(t)}{R(t)} = -\frac{4\pi G}{3}(\rho + 3p), \tag{41}$$

which we used with the cosmological constant to study inflation.

The quantum of the gravitational field, $G_{\mu\nu}$, called the graviton, has quantum spin $= 2$. Quantum fields with spin 2 quanta cannot be treated using the methods that work for vector quantum fields, like electromagnetism. That is why gravity and the graviton are not part of the Standard Model. On the other hand one can test for the existence of a graviton, which was discussed in Chapter 7, and we now review.

3.2.2. *Gravitational Quantum Theory*

The quantum field theory of gravity is very complicated. One of the most interesting aspects is possible gravitational radiation.

Consider the energy levels of the hydrogen atom, which are labelled by nl, with n the principal and l the angular momentum quantum number; and $l = 0, 1, 2$ called s, p, d, as in the figure.

The electron in the 2p drops to the 1s state emitting a photon with frequency $\omega_{2p-1s} = 6.27 \times 10^8$/sec or time to radiate 1.6×10^{-9} sec. Gravitational radiation from the 2p to the 1s state cannot take place as the spin of the graviton is 2, and the change in angular momentum between these two states is 1. The graviton radiation from the 3d to the 1s state is shown in the figure below.

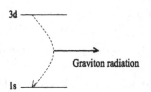

Since a graviton has spin=2 it cannot connect the electron in the 2p (l=1) to the 1s (l=0) state, but can connect the electron in the 3d (l=2) to the 1s state.

For graviton radiation from the 3d to the 1s state the frequency (see Weinberg [11] for derivation) $\omega_{3d-1s} = 2.5 \times 10^{-44}$/sec, so the time to radiate is 4×10^{43} sec, which is so large that there is no possibility to observe gravitational radiation from atoms.

3.2.3. *Gravity Waves and the EWPT*

Cosmological events, like inflation and the cosmological phase transitions — electroweak and quantum chromodynamic (to be discussed in the following chapter) —, can produce gravitational waves, much as dropping a rock in water can produce water waves.

The gravity wave energy density and amplitude, $h_c(f)$, with the frequency $f = \omega/(2\pi)$ have been calculated via Magnetohydrodynamic (MHD) Turbulence [12] using magnetic fields from the EWPT and QCDPT. The magnetic field energy density for gravity waves with wavelength λ as a function of $k = 2\pi/\lambda$ and time, $E_M(k,t)$, is given by v_A, the Alfvén velocity. With a magnetic field B and $w = \rho + p$, with ρ the energy density and p the pressure, which were in Friedmann's equation, $v_A = \frac{B}{\sqrt{4\pi w}}$.

Over a range of wave numbers k starting with k_0, the magnetic field energy density is

$$E_M(k,t) \simeq \epsilon^{2/3} k^{-5/3}, \tag{42}$$

with $\epsilon = (2/3)^{3/2} k_0 v_A^3$.

From $E_M(k,t)$ one can find the gravitational energy density and the amplitude of gravitational waves produced by B and v_A:

$$h_c(f) = 2 \times 10^{-14} \frac{100\,\text{GeV}}{T_*} \sqrt{\tau_T \omega B^{EWPT}(T_c)^4 H(\omega)}, \qquad (43)$$

where $B^{EWPT(T_c)}$ is the magnetic field created during the EWPT at the temperature $T = T_c$, τ_T is the time interval for the gravity wave creation during the EWPT, and $H(\omega)$ is obtained from the stress-energy tensor, T_{ij}, which we defined above. See Ref. [13] for the definition of $H(\omega)$ and references to earlier publications of gravity wave creation.

In the figure below $h_c(f)$ for the Electroweak Phase Transition (EWPT) at a time about 10^{-11} sec after the Big Bang, when the temperature of the universe was about the mass of the Higgs boson (about $125\,\text{GeV}$), is shown for $T_* = 80\,\text{GeV}$ (solid line), $100\,\text{GeV}$ (dashed line), and $150\,\text{GeV}$ (dash-dotted line), and compared to the amplitude expected to be detected by the future LISA detector (from Kahniashvili *et al.* [12]).

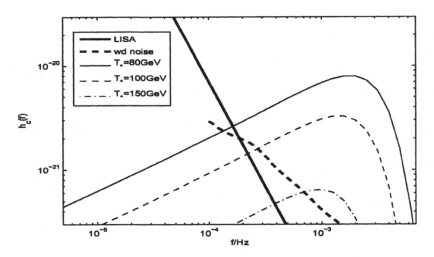

From this figure one can see that gravity waves produced during the EWPT could be detected by the LISA detector.

In Ref. [12] gravity waves produced by magnetic fields created during the QCDPT were also calculated. The magnetic field created by bubble collisions during the QCDPT [14] is

$$B^{QCDPT} \simeq 2.2 \times 10^{16} \text{ gauss}, \tag{44}$$

and $h_c(f)$ from the QCDPT was found to be too small to be detected by the LISA detector.

4. Problems

1. Prove that the function $\langle x|p_x \rangle = e^{ik_x x}$ where $k_x = p_x/\hbar$ is an eigenfunction of the p_x quantum operator.
2. Prove that $\phi(x) = ae^{ix}$, where a is a constant, is a solution to the EOM of Eq. (17).
3. Using Eqs. (37), (38), (39), (40) derive Friedmann's equation, Eq. (41).

References

[1] Alexander L. Fetter and John Dirk Walecka, *Quantum Theory of Many-Particle Systems*, published by McGraw-Hill, Inc.
[2] Leonard I. Schiff, *Quantum Mecanics*, published by McGraw-Hill, Inc.
[3] Ernest M. Henley, Mikkel B. Johnson, and Leonard S. Kisslinger, *Phys. Rev. D* **81**, 085035 (2010).
[4] John David Jackson, *Classical Electrodynamics*, published by John Wiley & Sons, Inc.
[5] Trevor Stevens, Mikkel B. Johnson, Leonard S. Kisslinger, Ernest M. Henley, W.-Y. Pauchy Hwang, and Matthias Burkardt, *Phys. Rev. D* **77**, 023501 (2008).
[6] Trevor Stevens and Mikkel B. Johnson, *Phys. Rev. D* **80**, 083011 (2009).
[7] Dario Grasso and Hector R. Rubinstein, *Phys. Rep.* **348**, 163 (2001).
[8] Dam Thanh Son, *Phys. Rev. D* **59**, 063008 (1999).
[9] Alexander G. Tevzadze, Leonard Kisslinger, Axel Brandenburg, and Tina Kahniashvili, *Astrophys. J.* **759**, 54 (2012).
[10] Edward W. Kolb and Michael S. Turner, *The Early Universe*, Addison-Wesley Publishing Co. (1990).
[11] Steven Weinberg, *Gravitation and Cosmology: Principles and Applications of the General Theory of Relativity*, John Wiley & Sons, Inc. (1972).
[12] Tina Kahniashvili, Leonard Kisslinger, and Trevor Stevens, *Phys. Rev. D* **81**, 023004 (2010).
[13] Grigol Gogoberidze, Tina Kahniashvili, and Arthur Kosowsky, *Phys. Rev. D* **81**, 083002 (2007).
[14] Leonard S. Kisslinger, *Phys. Rev. D* **68**, 043516 (2003).

Chapter 10

Quantum Chromodynamic Phase Transition (QCDPT)

In Chapter 9 the Electroweak Phase Transition (EWPT) was discussed. In the present chapter we explore the Quantum Chromodynamics (QCD) Phase Transition, QCDPT, which took place at about $t = 10^{-5}$ sec, when the temperature $T \simeq 150$ MeV.

As we saw in Chapter 9, if the phase transition is not first order, there is no critical temperature, T_c, and bubbles of the new phase do not form within the old phase. With a MSSM EW theory the EWPT is first order.

For many years lattice QCD calculations have found that the QCDPT is a crossover. See, e.g., Ref. [1]. It has now been shown that the QCDPT is a first order phase transition [2, 3], when the universe went from the quark–gluon plasma (QGP) to our universe of hadrons, $T_c \simeq 150$ MeV.

In this chapter we discuss various aspects of the first order QCDPT:

1. Magnetic field creation and CMBR correlations.
2. QCDPT magnetic fields as primordial seeds for magnetic fields in galaxies and clusters of galaxies.
3. Dark Energy, cosmological constant and the QCDPT quark condensate.
4. Creation of the QGP by relativistic heavy ion collisions (RHIC).

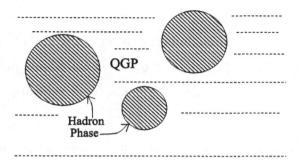

When $T \sim 150\,\text{MeV}$ the QCD phase transition starts: At time $t = 10^{-5}\,\text{s}$ bubbles of hadronic matter form within the quark–gluon plasma (QGP). By $t = 10^{-4}\,\text{s}$ quarks have condensed. The universe consists of protons, neutrons, etc., as well as leptons and photons. It is too hot for nuclei or atoms to form. Helium and other light nuclei form at about 1 minute after the Big Bang.

1. The QCD Phase Transition

With T_c the critical temperature for the QCDPT, for $T > T_c$ the universe was filled with a dense matter — a plasma with quarks and gluons — called the quark–gluon plasma (QGP). For $T < T_c$, our universe with protons and other hadrons was formed. During the time that $T = T_c$ bubbles of our universe nucleated within the QGP.

The QCDPT is a first order phase transition. Therefore it not only has a critical temperature, $T_c = 150\,\text{MeV}$, but there is latent heat created during the QCDPT.

THE QUARK CONDENSATE IS THE LATENT HEAT FOR THE QCDPT

QCD fields and particles (quarks) and the quark condensate:

$$q(x) = \text{quark field}$$
$$\bar{q}(x) = \text{antiquark field}$$
$$|T\rangle = \text{vacuum state temperature} = T$$

$$\langle T|\bar{q}(x)q(x)|T\rangle = \text{quark condensate}$$
$$= \text{vacuum expectation value of } \bar{q}(x)q(x)$$
$$\langle T|\bar{q}(x)q(x)|T\rangle = 0, T > T_c \text{ in quark gluon plasma phase}$$
$$\simeq -(0.23\,\text{GeV})^3, \ T < T_c \text{ in hadron phase.}$$

That is, with a first order QCD phase transition the quark condensate goes from zero to a finite value at the critical temperature (about 150 MeV).

Note that the quark condensate $\langle T|\bar{q}(x)q(x)|T\rangle$ is vacuum energy. As we discussed earlier Dark Energy is vacuum energy. Therefore it is possible that Dark Energy might be related to the QCDPT and the creation of the quark condensate. This is discussed below.

In the figure below are results of recent lattice gauge calculations (basic QCD theory computed on large computers). From this figure one can see the discontinuity in the quark condensate at the temperature 150 MeV. This shows that the QCDPT is first order, as was mentioned with references given previously.

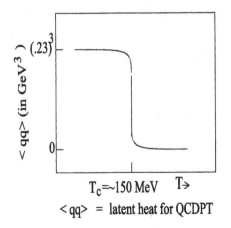

$T_c \approx 150\,\text{MeV}$ T→

$\langle qq \rangle$ = latent heat for QCDPT

T decreases from 300 MeV (t=10^{-5} s) to 100 MeV (t=10^{-4} s)

From the calculation shown in the figure one sees that the quark condensate $\langle \bar{q}q \rangle$ is $(0.23\,\text{GeV})^3$. As we discuss below, this is the local quark condensate, and a nonlocal quark is needed for an accurate estimate of vacuum energy.

1.1. *Magnetic Field Creation During the QCDPT*

As was discussed in Chapter 9 for the EWPT, in a first order phase transition bubbles are formed and nucleate, producing electromagnetic fields, and magnetic fields are formed during bubble collisions.

The surface of the bubbles is made of gluon fields, and after two QCDPT bubbles collide they form an interior wall also made of gluon fields, very similar to the collision and merger of two soap bubbles, as shown in the figure.

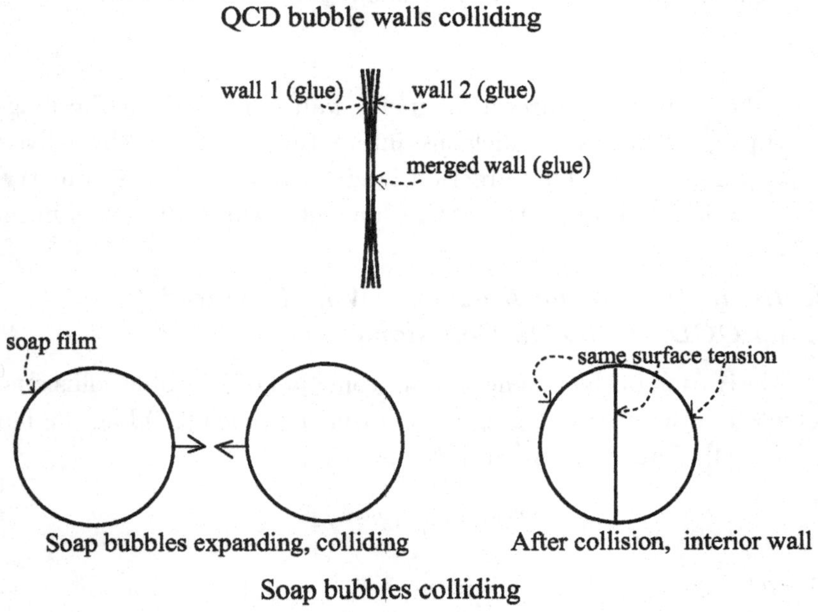

QCD bubble walls colliding

wall 1 (glue)   wall 2 (glue)

merged wall (glue)

soap film

same surface tension

Soap bubbles expanding, colliding   After collision, interior wall

Soap bubbles colliding

The quarks in the nucleons within the bubble interact with the gluonic wall, as is discussed next.

Magnetic field creation during QCDPT

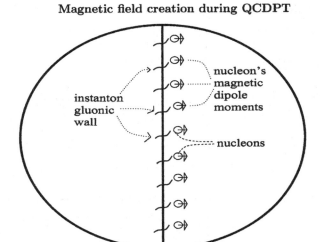

Oriented magnetic dipole moments form a magnetic wall

Gluonic wall decays in $\sim 10^{-8}$s. Magnetic wall lives on

As shown in the figure, a magnetic wall is formed by the magnetic dipole moments of nucleons interacting with the short-lived interior gluonic wall. The direction is given by a broken symmetry. The derivation and properties of this magnetic wall is discussed next.

1.2. *Derivation of the Magnetic Wall Produced by QCDPT Bubble Collisions*

In the derivation of the magnetic wall from QCDPT bubble collisions, illustrated in the preceeding figure, a purely gluonic QCD Lagrangian was used [4]. This Lagrangian $\mathcal{L}^{glue}$ is

$$\mathcal{L}^{glue} = G^a_{\mu\nu}G^{\mu\nu a}/4,$$

with $G^{\mu\nu a}$ the color field tensor, with $\mu, \nu = 0, 1, 2, 3$ and $a, b, \ldots = 1, 2, 3$ Dirac indices and color indices, respectively. From $\mathcal{L}^{glue}$ one can derive the gluonic wall, which was shown in the figure.

The magnetic wall is formed by the interaction of the nucleons with the gluonic wall, with the electromagnetic interaction

Lagrangian

$$\mathcal{L}^{int} = -e\bar{\Psi}\gamma^{\mu}A_{\mu}^{em}\Psi, \tag{1}$$

where Ψ is the nucleon field operator and γ^{μ} is the Dirac operator, a 4×4 matrix defined by

$$\gamma^{\mu}\gamma^{\nu} + \gamma^{\nu}\gamma^{\mu} = 2g^{\mu\nu},$$

where $g^{\mu\nu}$ is the metric tensor ($g^{00} = 1$, $g^{ii} = -1$, $g^{\mu\nu} = 0$, $\mu \neq \nu$) and A_{μ}^{em} is the electromagnetic field that we discussed previously. This leads to the electromagnetic interaction with the nucleon's magnetic dipole moment given by

$$\nu^{int} = \frac{e}{2M_n}\bar{\Psi}\sigma_{\mu\nu}\gamma_5\Psi F^{\mu\nu}, \tag{2}$$

where $\gamma_5 = i\gamma^0\gamma^1\gamma^2\gamma^3$, $\sigma_{\mu\nu} = i[\gamma_{\mu}\gamma_{\nu} - \gamma_{\nu}\gamma_{\mu}]/2$, and $F^{\mu\nu}$ is defined in terms of the electromagnetic field, A^{μ} by

$$F^{\mu\nu} = \partial^{\mu}A^{\nu} - \partial^{\nu}A^{\mu}.$$

For the wall oriented in the x, y-direction with thickness $\rho B_z \equiv B_W = F^{21}$, which can be shown to be (ρ = wall thickness)

$$B_z \simeq \frac{1}{\rho\Lambda_{QCD}}\frac{e}{2M_n}\langle\bar{\Psi}\sigma_{21}\gamma_5\Psi\rangle, \tag{3}$$

where the momentum of the fermions (nucleons) in the plane of the wall is $\Lambda_{QCD} = T_{QCD} \simeq 150\,\text{MeV}$ and in the z-direction. From the evaluation of $\langle\bar{\Psi}\sigma_{21}\gamma_5\Psi\rangle$ [4] one finds

$$B_W \simeq 10^{17}\,\text{gauss}, \tag{4}$$

and therefore, at the end of the QCDPT ($t \simeq 10^{-4}\,\text{s}$), the magnetic field produced by bubble collisions in coordinate space is

$$B_W(\vec{x}) = B_W e^{-b^2(x^2+y^2)}e^{-M_n^2 z^2}, \tag{5}$$

or in momentum space

$$B_W(\vec{k}) = \frac{B_W}{2\sqrt{2}b^2 M_n}e^{-(k_x^2+k_y^2)/4b^2}e^{-k_z^2/4M_n^2}, \tag{6}$$

where b^{-1} is of the scale of the horizon size, $b^{-1} = d_H \simeq$ a few km, while $M_n^{-1} \simeq 0.2$ fm, with 1 fm $= 10^{-13}$ cm, is the thickness of the wall. Therefore, although B_W is very large, since the wall occupies a very small volume of the universe, such a structure is compatible with galaxy structure.

As will be discussed in a section that follows, from the magnitude of B_W given in Eq. (4), the magnetic field created during the QCDPT could be the primary seed for galaxy and galaxy cluster magnetic fields, a problem that many theorists have tried to solve.

Next effects of $B_W(\vec{k})$ on CMBR will be discussed.

1.2.1. *Polarization Correlations from the QCDPT Magnetic Wall*

In our discussion of the CMBR only temperature–temperature correlations, C_l^{TT} were treated in detail. The possible effects on the CMBR from the QCDPT magnetic field are magnetic field–magnetic field correlations, C_l^{BB}. The $B_z B_z$ correlation is

$$\langle B_z(\vec{k}, \eta) B_z(\vec{k}', \eta) \rangle \simeq B_W^2 \delta(k_x - k_x') \delta(k_y - k_y') \langle e^{-k_z^2/4M_n^2} e^{-k_z'^2/4M_n^2} \rangle$$

$$\simeq B_W^2 d_H e^{-k_z^2/4M_n^2} \delta(\vec{k} - \vec{k}'). \tag{7}$$

This gives for the polarization power spectrum

$$C_l^{BB} = \frac{(l+1)(l+2)}{\pi} B_W^2 d_H \int dk k^2 \frac{j_l^2[k(\Delta\eta)]}{k^2(\Delta\eta)^2}, \tag{8}$$

where η is the conformal time and the conformal time integral over the visibility function has been carried out and $\Delta\eta$ is the conformal time width at the last scattering, the time when the CMBR was released.

Using the parameters $M_n \Delta\eta = 1.5 \times 10^{39}$ (from Refs. [4, 5]), $d_H = 0.37 \times 10^{24}$ GeV^{-1}, and $B_W = 1.0 \times 10^{17}$ gauss,

$$C_l^{BB} \simeq 4.25 \times 10^{-8} l^2 \tag{9}$$

The result for the B-type power spectrum is shown in Fig. 10.1 by the solid line. There have been a number of investigations of the

polarization predicted by inflationary models [5] which show that the
B-type polarization in inflationary models $l(l+1)C_l^{BB}$ peaks at l-
values about 100. The curve shown by small circles is a string model
normalized at $l = 100$ to our magnetic wall value, and with the same
normalization the dashed curve gives a typical inflationary model
result. One can see that for $l \simeq 1000$ the values of C_l^{BB} predicted by
our magnetic wall picture exceed those of inflationary and topological
models, and the l-dependence is quite different.

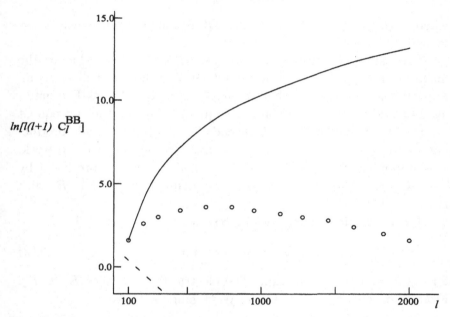

Fig. 10.2. B-type power spectrum in magnetic wall model (solid), string (circles),
inflationary (dashed).

2. QCDPT and Magnetic Fields in Galaxies and Galaxy Clusters

In this section the Primary Magnetic Field (PMF) as seeds for galaxy
and galaxy cluster magnetic fields will be reviewed.

As was discussed in Section 1.2, the magnetic field created via
QCDPT bubble collisions is, with the normal to the interier wall in

the z-direction,

$$B_z(x) = B_*^{QCD} e^{-b^2(x^2+y^2)} e^{-M_n^2 z^2}, \tag{10}$$

where $b^{-1} = d_H \simeq$ a few km = horizon size at the end of the QCDPT $(t \simeq 10^{-4}\,\text{s})$ and $M_n^{-1} = 0.2\,\text{fm}$.

B_*^{QCD} the magnitude of the magnetic field within the wall of thickness ζ, is

$$B_*^{QCD} \simeq \frac{1}{\zeta \Lambda_{\text{QCD}}} \frac{e}{2M_n} \times \langle \bar{\Psi}\sigma_{21}\gamma_5\Psi\rangle, \tag{11}$$

where $\Lambda_{\text{QCD}} \simeq 0.15\,\text{GeV}$ is the QCD momentum scale, $\gamma_5 = i\gamma^0\gamma^1\gamma^2\gamma^3$, and $\sigma_{21} = i\gamma_2\gamma_1 = i\gamma^2\gamma^1$.

The asterisk indicates a reference to the *initial* value of the magnetic field at the time of the QCDPT. Magnetic helicity is an important characteristic that strongly influences the PMF dynamics. Magnetic helicity is a conserved quantity during the subsequent evolution past the QCDPT. This leads to an inverse cascade producing magnetic fields at progressively larger scales. For this to work, it is important to know the magnetic helicity that is produced by the QCDPT. The magnetic helicity is defined as $\int d^3 x A \cdot B$, with $B = \nabla \times A$.

The magnetic helicity density $\mathcal{H}_M$ is

$$\mathcal{H}_M = A \cdot B = A_z B_z, \tag{12}$$

for a PMF in the z-direction. The electric field satisfies $E_z \simeq B_z$. From Maxwell's equations in the Weyl gauge we have

$$E = -\frac{1}{c}\frac{\partial A}{\partial t} \quad \text{or} \quad A_z \simeq -E_z \tau, \tag{13}$$

where $\tau \simeq 1/\Lambda_{\text{QCD}}$ is the time scale for the QCDPT. From Eqs. (12) and (13) one finds

$$\mathcal{H}_{M,*}^{\text{QCD}} \simeq B_z^2/\Lambda_{\text{QCD}}$$
$$\simeq (0.22 \times 10^{17}\,\text{G})^2/(0.15\,\text{GeV}), \tag{14}$$

where we have assumed statistical homogeneity, so the result is gauge-independent.

The coupling between the initial QCDPT-generated magnetic field and the plasma uses the basic MHD equations for an incompressible conducting fluid [7]

$$\left[\frac{\partial}{\partial \eta} + (v \cdot \nabla) - \nu \nabla^2\right] v = (b \cdot \nabla)b - \nabla p + f_K, \qquad (15)$$

$$\left[\frac{\partial}{\partial \eta} + (v \cdot \nabla) - \lambda \nabla^2\right] b = (b \cdot \nabla)v + \nabla \times f_M, \qquad (16)$$

where $\eta = t/R(t)$ is the conformal time, $v(x, \eta)$ is the fluid velocity, $b(x, \eta) \equiv B(x, \eta)/\sqrt{4\pi w}$ is the normalized magnetic field, $f_K(x, \eta)$ and $f_M(x, \eta)$ are external forces driving the flow and the magnetic field ($f_K = f_M = 0$ for the results presented below, but $f_M \neq 0$ for producing initial conditions), ν is the comoving viscosity of the fluid, λ is the comoving resistivity, $w = \rho + p$ with ρ the energy density, and p the pressure of the plasma.

The Fourier transform of the PMF two point correlation function is

$$\langle b_i^*(\vec{k}, \eta) b_j(\vec{k}', \eta + \tau) \rangle = (2\pi)^3 \delta(\vec{k} - \vec{k}') F_{ij}^M(\vec{k}, \tau) f[\kappa(k), \tau], \qquad (17)$$

The spectral correlation tensor is

$$\frac{F_{ij}^M(\vec{k}, \tau)}{(2\pi)^3} = P_{ij}(\vec{k}) \frac{E_M(k, \tau)}{4\pi k^2} + i\varepsilon_{ijl} k_l \frac{H_M(k, \tau)}{8\pi k^2}. \qquad (18)$$

Here, $P_{ij}(\vec{k}) = \delta_{ij} - k_i k_j/k^2$, $k = |\vec{k}|$, ε_{ijl} is the totally antisymmetric tensor, and $\kappa(k)$ is an autocorrelation function that determines the characteristic function $f[\kappa(k), \tau]$ describing the temporal decorrelation of turbulent fluctuations. The function $H_M(k, \eta)$ is the magnetic helicity spectrum. Note that $E_M(k) = k^2 P_B(k)/\pi^2$, where $P_B(k)$ is the magnetic field power spectrum.

The power spectra of magnetic energy $E_M(k, \eta)$ and magnetic helicity $H_M(k, \eta)$ are related to magnetic energy density and helicity density through $\mathcal{E}_M(\eta) = \int_0^\infty dk E_M(k, \eta)$ and $\mathcal{H}_M(\eta) = \int_0^\infty dk H_M(k, \eta)$, respectively. The magnetic correlation length,

$$\xi_M(\eta) = [2\pi/\mathcal{E}_M(\eta)] \int_0^\infty dk \, k^{-1} E_M(k, \eta), \qquad (19)$$

corresponds to the largest eddy length scale.

Carrying out the rather complicated calculations using the equations above, it was shown [6] that

1. The correlation length is approximately $10\,\mathrm{kpc}$.
2. The amplitude of the magnetic field is approximately 7×10^{-12} gauss, which is consistent with observed magnetic fields (see, e.g., Ref. [8]).

Therefore, the puzzle of galactic and galactic cluster magnetic fields might have been solved.

3. Dark Energy and the QCDPT Quark Condensate

As was discussed in Chapter 7 when inflation was derived, Einstein introduced a cosmological constant, Λ in his equation of General Relativity to get a static universe. In 1915 Einstein added a new term, $\Lambda g_{\mu\nu}$ to the Einstein tensor [9], so the Einstein's equation in tensorial form became

$$\mathcal{R}_{\mu\nu} - \frac{1}{2}\mathcal{R}g_{\mu\nu} + \Lambda g_{\mu\nu} = 8\pi G T_{\mu\nu}. \tag{20}$$

where $\mathcal{R}_{\mu\nu}$, the Ricci tensor, $\mathcal{R}$, the trace of the Ricci tensor, and the energy-momentum tensor $T_{\mu\nu}$ were defined in Chapter 7, with $-T_{00} = $ the universe's energy density, ρ.

Note that the term $\Lambda g_{\mu\nu}$ in Eq. (20) corresponds to adding a vacuum term to $T_{\mu\nu}$,

$$T_{\mu\nu}(vac) = \rho_\Lambda g_{\mu\nu}. \tag{21}$$

Therefore, the cosmological constant Λ is related to the vacuum energy density, ρ_Λ (previously known as ρ_{VAC}) by

$$\Lambda = 8\pi G\rho_\Lambda. \tag{22}$$

Recently it was shown that the vacuum energy density ρ_Λ can be derived from the quark condensate, which is the latent heat for the QCDPT as discussed earlier; however, as shown in Ref. [10] the nonlocal quark constant is necessary for an accurate relationship.

Urban and Zhitnitsky [11] introduced a model in which the cosmological vacuum energy density ρ_Λ can be expressed in terms of

QCD parameters for $N_f = 2$ light flavors as follows

$$\rho_\Lambda = c \frac{2HN_f}{m_{\eta'}} |m_q \langle 0| : \bar{q}(0)q(0) : |0\rangle|, \tag{23}$$

where $m_{\eta'}$ is the mass of the η' meson, m_q is the current quark mass and $c \simeq c_{grav..}$

$c_{grav.}$ is defined as the relation between the size L of the manifold we live in, and the Hubble constant H, $L = (c_{grav.}H_0)^{-1}$ (H_0 is the Hubble constant today). One can define this size of the manifold as $L \simeq 17H_0^{-1}$ where $H_0 = 2.1 \times 10^{-42} \times h\,\text{GeV}$ and $h = 0.71$. Therefore, one can explicitly obtain an estimate for the linear length L of the torus, and then obtain the value of $c_{grav.}$ with $c_{grav.} = 0.0588$.

3.1. *Local Quark Condensate, Current Quark Mass, ρ_Λ*

The quark condensate is the nonperturbative scalar part of the quark propagator:

$$S_q^{NP}(x) = -\frac{1}{12} \langle 0| : \bar{q}(x)q(0) : |0\rangle. \tag{24}$$

For short distances, the Taylor expansion of $\langle 0| : \bar{q}(x)q(0) : |0\rangle$ can be written as

$$\langle 0| : \bar{q}(x)q(0) : |0\rangle = \langle 0| : \bar{q}(0)q(0) : |0\rangle$$
$$-\frac{x^2}{4} \langle 0| : \bar{q}(0)[ig_s\sigma G(0)]g(0) : |0\rangle + \cdots \tag{25}$$

In Eq. (25) the vacuum expectation values in the expansion are the local quark condensate, the quark–gluon mixed condensate, and so forth, but the Taylor expansion is valid only for small x.

The current quark masses of the light u and d quarks are needed. Estimates of these masses in the Particle Data Physics booklet are

$$1.7 < m_u < 3.3\,\text{MeV}$$
$$4.1 < m_d < 5.8\,\text{MeV}. \tag{26}$$

From this we estimate that the current quark mass is

$$m_q \simeq 4.0\,\text{MeV}. \tag{27}$$

Using the standard value of the local quark condensate, $\langle \bar{q}(0)q(0) \rangle = -(0.23\,\text{GeV})^3$, one obtains

$$\rho_\Lambda^{theory} \simeq (3.6 \times 10^{-3}\,\text{eV})^4, \tag{28}$$

while the value observed [12] is

$$\rho_\Lambda^{observed} \simeq (2.3 \times 10^{-3}\,\text{eV})^4. \tag{29}$$

Although the theoretical and observed values are similar, they still differ by

$$\rho_\Lambda^{theory}/\rho_\Lambda^{observed} \simeq 6.0.$$

3.2. *Nonlocal Quark Condensate and* ρ_Λ

As mentioned above, the Taylor expansion $\langle 0| : \bar{q}(x)q(0) : |0\rangle = \langle 0| : \bar{q}(0)\,q(0) : |0\rangle - \frac{x^2}{4}\langle 0| : \bar{q}(0)[ig_s\sigma G(0)]q(0) : |0\rangle + \cdots$ does not work except for very small x.

Therefore the nonlocal quark condensate derived from the quark distribution function (see Refs. [13, 14]) is used, with the form given in Ref. [14]:

$$\langle 0| : \bar{q}(x)q(0) : |0\rangle = g(x^2)\langle 0| : \bar{q}(0)q(0) : |0\rangle, \tag{30}$$

with

$$g(x) = \frac{1}{(1 + \lambda^2 x^2/8)^2}. \tag{31}$$

The value of λ^2 estimated in Ref. [15] is $\lambda^2 \simeq 0.8\,\text{GeV}^2$. Using $1/\Lambda_{QCD}$ as the length scale, or $x^2 = (1/0.2\,\text{GeV})^2$, one obtains

$$g(1/\Lambda_{QCD}) = \frac{1}{2.25^2} = \frac{1}{6.25}. \tag{32}$$

From this we obtain

$$\langle 0| : \bar{q}(x)q(0) : |0\rangle = \frac{1}{6.25}\langle 0| : \bar{q}(0)q(0) : |0\rangle, \tag{33}$$

and

$$\rho_\Lambda^{nonlocal\ theory} \simeq \frac{1}{6}(3.6 \times 10^{-3}\,\text{eV})^4$$

$$= (2.3 \times 10^{-3}\,\text{eV})^4 \simeq \rho_\Lambda^{observed}. \qquad (34)$$

Therefore, using the modification of the quark condensate via the nonlocal condensate, one obtains excellent agreement between the theoretical and observed cosmological constants.

3.3. *Time (z) Dependence of Dark Energy Density*

In the previous subsection it was shown that the present magnitude of the Dark Energy density, ρ_Λ, which is vacuum energy, is given by the nonlocal quark condensate.

The quark condensate was created during the QCDPT, reaching its present value at about 10^{-4} sec. Note that in most models of inflation using a quintessence field Dark Energy went to zero at about 10^{-32} sec. If the model given in the previous subsection is correct, $\rho_\Lambda = \rho_\Lambda(t)$, or in astrophysical notation Dark Energy density is a function of z. The authors of Ref. [10] point out that this time and temperature dependence could be found by solving the temperature dependent Dyson–Schwinger equations, which is a project underway. Astrophysical observations might be able to test this z-dependence by measuring ρ_Λ in galaxies at large distance (large z).

4. Creation of the QGP by Relativistic Heavy Ion Collisions (RHIC)

As was discussed in detail in Section 1 of this chapter, at time $t = 10^{-5}$ sec, when the universe composed of a dense plasma of quarks and gluons (QGP), reached the critical temperature $T_c \simeq 150\text{MeV}$, the quarks and gluons condensed to form hadrons, which later became atomic nuclei and atoms.

For some decades there has been a great deal of interest in creating the QGP via relativistic heavy ion collisions (RHIC). An accelerator at Brookhaven National Laboratory (BNL) in the USA was designed with beams of heavy ions, such as Pb, collided, and the energy converted to heat reaches a temperature of greater than

150 MeV in at least part of the ions. The QGP should be created. More recently at the Large Hadron Collider (LHC) in Europe a preliminary RHIC program with greater energy than BNL was started. Both BNL-RHIC and LHC have carried our proton–proton (p–p) collisions as a preliminary to studies of RHIC with nucleus–nucleus (A–A) collisions.

A crutial question is how does one prove that the QGP has been created. One possibility is by creating heavy quark mesons with active glue, called hybrid mesons. We will see that experiments with p–p collisions producing heavy mesons are consistent with certain mesons being hybrids, and not ordinary quark–antiquark mesons. As a background for discussing the heavy quark state production in p–p collisions, Ref. [16], mixed hybrid heavy quark mesons will be reviewed.

4.1. *Mixed Hybrid Heavy Quark Mesons*

The heavy quarks that are being considered are charm (c) and bottom (b), with the mesons that have charm quarks called Charmonium, and those with bottom quarks called Upsilon states. The Charmonium and Upsilon (nS) states which are being studied are shown in the Fig. 10.2, with energy given in MeV.

All of these states have spin 1. Calling q^a a quark operator with color a, the operator that creates a standard heavy quark meson with spin 1 is

$$J_H^\mu = \bar{q}^a \gamma^\mu \gamma_5 q^a, \tag{35}$$

with γ^ν, γ_5 Dirac matrices defined above, and there is a sum over color a. One can show that J_H^μ creates an ordinary meson with quantum spin 1: $J_H^\mu | \ \rangle = |\bar{q}q, \text{spin} = 1, s_z = \mu\rangle$.

The operator that creates a heavy hybrid-meson state is more complicated. It is

$$J_{HH\mu} = \bar{q}^a \gamma^\nu \gamma_5 \frac{\lambda_{ab}^n}{2} \tilde{G}_{\mu\nu}^m q^b, \tag{36}$$

with λ_{ab}^n an operator that mixes color and $\tilde{G}_{\mu\nu}^m$ is the gluon field operator. There are sums over a, b, n. One can show that $J_{HH\mu}$ creates a

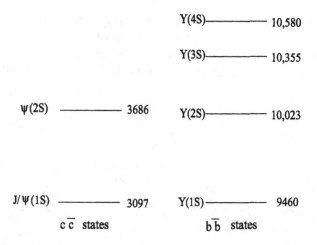

Fig. 10.2.   Lowest energy Charmonium and Upsilon states.

hybrid-meson state, $J_{HH\mu}|\ \rangle = |\bar{q}qg,\ spin = 1,\ s_z = \mu\rangle$, where $\bar{q}qg$ means that a quark, antiquark and gluon combine colors to produce a colorless meson with active glue. These operators will not be used, but are given to define heavy meson and heavy hybrid-meson states.

In Ref. [17] it was shown (see Fig. 10.2) that the $J/\Psi(1S)$ is an ordinary $c\bar{c}$ Chaxmonium state, and that the $\Upsilon(1S), (2S), (4S)$ are ordinary $b\bar{b}$ Upsilon states, while the $\Psi(2S)$ and $\Upsilon(3S)$ were found to be approximately 50% a heavy quark meson and 50% a hybrid:

$$|\Psi(2s)\rangle = -0.7|c\bar{c}(2S)\rangle + \sqrt{1 - 0.5}|c\bar{c}g(2S)$$
$$|\Upsilon(3S)\rangle = -0.7|b\bar{b}(3S)\rangle + \sqrt{1 - 0.5}|b\bar{b}g(3S)\rangle. \qquad (37)$$

Since the goal is to test the creation of the QGP, a plasma with active glue, this suggested that the creation of the $\Psi(2S)$ and $\Upsilon(3S)$ might provide such a test. First the creation of heavy meson states in p–p collisions is reviewed.

4.2. *Heavy Meson State Production in p–p Collisions*

The color octet model was used to estimate the production of heavy meson states in p–p collisions.

The cross sections for heavy state (H) production in the color octet model are based on the cross sections obtained from the matrix elements for quark–antiquark and gluon–gluon octet fusion to a hadron H, $\sigma_{q\bar{q}\to H(\lambda)}$ and $\sigma_{gg\to H(\lambda)}$, with λ the helicity, as illustrated in the figure.

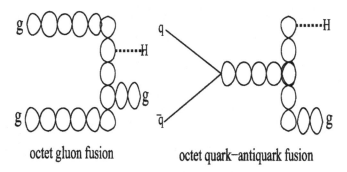

octet gluon fusion   octet quark–antiquark fusion

Fig. 10.3.   Gluon and quark–antiquark color octet fusion producing hadron H.

For the differential cross sections one needs the variables x and $y(x)$, which is defined in terms of the energy E and z-component of momentum p_z:

$$E = \sqrt{m^2 + p_z^2}$$

$$p_z = \frac{\sqrt{s}}{2}\left(x - \frac{a}{x}\right) \qquad (38)$$

$$y(x) = \frac{1}{2}ln\left(\frac{E + p_z}{E - p_z}\right),$$

where $s = E^2$ and $a = 4m^2/s$, with $m =$ the quark mass ($m =$ 1.5 GeV and 5 GeV for Charmonium and Bottomonium). The variable $y(x)$ is called rapidity.

To obtain the production cross sections one needs to multiply the quark–antiquark or gluon–gluon octet cross sections by the quark–antiquark or gluon parton distribution functions, giving

$$\sigma_{pp\to\Phi(\lambda)} = \int_a^1 \frac{dx}{x}[f_q(x, 2m)f_{\bar{q}}(a/x, 2m)\sigma_{q\bar{q}\to H(\lambda)}$$

$$+ f_g(x, 2m)f_g(a/x, 2m)\sigma_{gg\to H(\lambda)}], \qquad (39)$$

where $f_g(x, 2m)$, $f_q(x, 2m)$ are the gluonic and quark distribution functions, which have been tabulated from the analysis of many experiments. λ is the helicity, which can be 0 or 1 for the production of the spin one states that are being considered.

The differential rapidity distributions for $\lambda = 0, 1$ are given by

$$\frac{d\sigma_{pp \to \Phi(\lambda=0)}}{dy} = A_\Phi \frac{1}{x(y)} f_g(x(y), 2m) f_g\left(\frac{a}{x(y)}, 2m\right) \frac{dx}{dy}, \qquad (40)$$

$$\frac{d\sigma_{pp \to \Phi(\lambda=1)}}{dy} = A_\Phi \frac{1}{x(y)} \left[f_g(x(y), 2m) f_g\left(\frac{a}{x(y)}, 2m\right) \right.$$

$$+ 0.613 \left(f_d(x(y), 2m) f_{\bar{d}}\left(\frac{a}{x(y)}, 2m\right) \right.$$

$$\left. + f_u(x(y), 2m) f_{\bar{u}}\left(\frac{a}{x(y)}, 2m\right) \right) \left. \right] \frac{dx}{dy}, \qquad (41)$$

where the parameter A_Φ depends on the octet marix elements, the quark mass, and the energy. For the calculations presented here the ones needed are as follows. For E $= 200\,$GeV (BNL) $A_\Phi = 7.9 \times 10^{-4}\,$nb for Charmonium production and $2.13 \times 10^{-5}\,$nb for Upsilon production. The quantity nb $=$ nanobarn $= 10^{-9}$ barn $= 10^{-37}\,$m^2. For E $= 38.8\,$GeV (Fermilab) $A_\Phi = 5.66 \times 10^{-4}\,$nb for Upsilon production. For E $= 2.76\,$TeV $= 2{,}760\,$GeV, $A_\Phi = 1.12 \times 10^{-7}\,$nb for Upsilon production.

Since the absolute values of the cross sections for the $\Psi(nS)$ and $\Upsilon(nS)$ are difficult to calculate, the most important theoretical results that will be shown below are ratios of cross sections.

4.3. *Charmonium Production Via Unpolarized p–p Collisions at E = 200 GeV at BNL-RHIC*

The results for J/Ψ production are not shown, since the $J/\Psi(1S)$ is a standard $c\bar{c}$ meson.

The results for dσ/dy for $\Psi'(2S)$ production, shown in Fig. 10.4 labeled $\Psi'(2S)(a)$ are obtained by using for the standard Charmonium model, while the results labeled $\Psi'(2S)(b)$ is obtained by using the matrix element derived using the result that the $\Psi(2S)$ is approximately 50% a hybrid with the enhancement at least a factor of π.

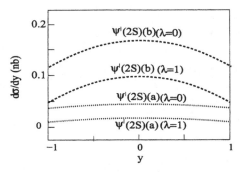

Fig. 10.4. $d\sigma/dy$ for $Q = 3\,\text{GeV}$, $E = 200\,\text{GeV}$ unpolarized p–p collisions producing $\Psi'(2S)$ with $\lambda = 1$, $\lambda = 0$.

Since the production of the $\Psi(2S)$ state at BNL-RHIC has not yet been acheived, this is a prediction of the mixed heavy quark hybrid theory. The results for polarized p–p collisions for $E = 38.8\,\text{GeV}$ at Fermilab are similar, and are not shown.

4.4. *Upsilon Production Via Unpolarized p–p Collisions at 2.76 TeV at LHC-CMS*

The cross sections for $\Upsilon(nS)$ state production in p–p collisions have been measured at 2.76 TeV at the LHC-CMS [18].

The ratios of cross sections in the standard $b\bar{b}$ are

$$\sigma(\Upsilon(2S))/\sigma(\Upsilon(1S)) \simeq 0.27 \text{ standard}$$
$$\sigma(\Upsilon(3S))/\sigma(\Upsilon(1S)) \simeq 0.04 \text{ standard, giving}$$
$$\frac{\sigma(\Upsilon(2S)) + \sigma(\Upsilon(3S))}{\sigma(\Upsilon(1S))} \simeq 0.31 \text{ standard} \tag{42}$$

On the other hand, in the mixed hybrid theory with the $\Upsilon(3S)$ about 50% hybrid [17], there is a factor of $\pi^2/4$ in the matrix element, and therefore a factor of about 2.45 for the $\Upsilon(3S)$ cross section compared to the standard model. This results in the estimate

$$\frac{\sigma(\Upsilon(2S)) + \sigma(\Upsilon(3S))}{\sigma(\Upsilon(1S))} = 0.52, \tag{43}$$

compared to the LHC-CMS result [18] that this ratio is $0.78^{0.16}_{-0.14} \pm$ 0.02, which is consistent within experimental and theoretical errors with the mixed-hybrid theory, and not consistent with the standard model.

4.5. *Upsilon Production Via Unpolarized p–p Collisions at 38.8 GeV (Fermilab)*

The study of the 38.8 GeV Upsilon production is similar to the preceding one for the LHC-CMS 2.76 TeV experiments.

The result for the $\sigma(\Upsilon(3S))/\sigma(\Upsilon(1S))$ expected at 38.8 GeV in the standard model and the mixed-hybrid theory:

$$\frac{\sigma(\Upsilon(3S))}{\sigma(\Upsilon(1S))} \simeq 0.04 \text{ standard}$$

$$\frac{\sigma(\Upsilon(3S))}{\sigma(\Upsilon(1S))} \simeq 0.147 - 0.22 \text{ hybrid} \tag{44}$$

compared to the experimental result [19] of about 0.12 to 0.16.

4.6. *Upsilon Production in p–p Collisions at 7 TeV*

Recently here has been a publication by the CMS Collaboration at the LHC giving $\Upsilon(nS)$ cross sections at 7.0 TeV [20], with a much larger data set, in which both the $\Upsilon(2S)$ and $\Upsilon(3S)$ ratios to the $\Upsilon(1S)$ cross sections were measured. Thus one is not restricted to the ratio given in Eq. (43). The results given next have recently been published in Ref. [21]

The ratios of the $\Upsilon(2S)$ to $\Upsilon(1S)$ cross section at 7.0 TeV in both the standard model and the mixed-hybrid theory are

$$\sigma(\Upsilon(2S))/\sigma(\Upsilon(1S)) \simeq 0.27, \tag{45}$$

while

$$\sigma(\Upsilon(3S))/\sigma(\Upsilon(1S)) \simeq 0.04 \text{ standard}$$
$$\sigma(\Upsilon(3S))/\sigma(\Upsilon(1S)) \simeq 0.1 \text{ mixed hybrid.} \tag{46}$$

The CMS results at 7.0 TeV are [20]

$$\sigma(\Upsilon(2S))/\sigma(\Upsilon(1S)) \simeq 0.26 \pm 0.02 \pm 0.04 \text{ CMS}$$
$$\sigma(\Upsilon(3S))/\sigma(\Upsilon(1S)) \simeq 0.14 \pm 0.01 \pm 0.02 \text{ CMS.} \tag{47}$$

Therefore one sees that the CMS results for the $\sigma(\Upsilon(3S))/$ $\sigma(\Upsilon(1S))$ ratio are in disagreement with the standard quark model, but agree within errors with the mixed hybrid theory.

4.7. *Upsilon Production in p–p Collisions for Forward Rapidities at the LHC*

The studies of heavy meson production at the LHC discussed above were for rapidites in the range $-1.0 \leq y \leq 1.0$, the range for the CMS (Compact Muon Solenoid) detector. In anticipation of a future muon spectrometer, calculations of the differential rapidity cross sections for the range $2.5 \leq y \leq 4.0$ for $\Upsilon(1S)$ and $\Upsilon(3S)$ states were recently published [22].

The theory was given above. The new parameter needed is $A_\Upsilon = 1.73 \times 10^{-8}\,\mathrm{nb}$ for E $= 7.0\,\mathrm{TeV}$, while $A_\Upsilon = 1.12 \times 10^{-7}\,\mathrm{nb}$ for $2.76\,\mathrm{TeV}$, as was mentioned above. The results are shown in the figures below.

These results for the rapidity dependence of dσ/dy will be useful in the forward rapidity Bottomonia measurements with ALICE detector at LHC. Also, the ratios of the production of $\Upsilon(3S)$ to $\Upsilon(1S)$ will be both a test of the validity of the mixed hybrid theory, and should also be a guide for ALICE.

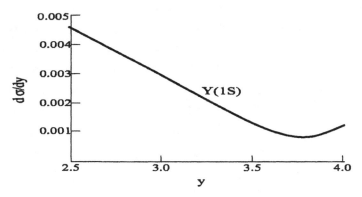

Fig. 10.5.   dσ/dy for p–p collisions at $\sqrt{s} = 2.76\,\mathrm{TeV}$ producing $\Upsilon(1S)$.

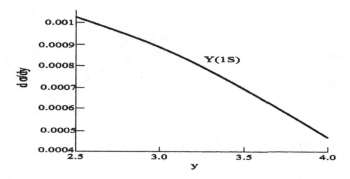

Fig. 10.6. $d\sigma/dy$ for p–p collisions at $\sqrt{s} = 7\,\text{TeV}$ producing $\Upsilon(1S)$.

5. Heavy Quark Meson Production in A–A Collisions at $\sqrt{s} = 200\,\text{GeV}$

In a study of RHIC at BNL, calculations of Lead–Lead (Pb–Pb) and Gold–Gold (Au–Au) collisions to produce Charmonium and Upsilon states as a test of the creation of the QGP are in progress [23].

The differential rapidity cross section for the production of a heavy quark state with helicity $\lambda = 0$ in the color octet model in A–A collisions is given by

$$\frac{d\sigma_{AA \to \Phi(\lambda=0)}}{dy} = R_{AA} N_{bin}^{AA} \langle \frac{d\sigma_{pp \to \Phi(\lambda=0)}}{dy} \rangle S_{\Phi}, \qquad (48)$$

where R_{AA} is the nuclear modification factor (see Ref. [24] for the definition and discussion), N_{bin}^{AA} is the number of binary collisions in the AA collision, S_{Φ} is the dissociation factor after the state Φ (a Charmonium or Bottomonium state) is formed, and $\langle \frac{d\sigma_{pp \to \Phi(\lambda=0)}}{dy} \rangle$ is the differential rapidity cross section for Φ production via nucleon-nucleon collisions in the nuclear medium. See Refs. [25, 26].

Experimental studies show that for $\sqrt{s} = 200\,\text{GeV}$, $R_{AA} \simeq 0.5$ both for Cu–Cu and Au–Au. The number of binary collisions are $N_{bin}^{AA} = 51.5$ for Cu–Cu and 258 for Au–Au. It is assumed that $S_{\Phi} = 1$.

From this the differential rapidity cross sections for J/Ψ and $\Upsilon(1S)$ production via Cu–Cu and Au–Au collisions at RHIC (E = 200 GeV) shown in the following figures were obtained.

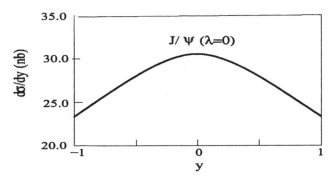

Fig. 10.7.   dσ/dy for Q = 3 GeV, E = 200 GeV, Cu–Cu collisions producing J/Ψ with $\lambda = 0$.

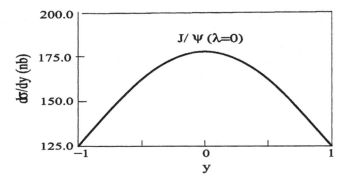

Fig. 10.8.   dσ/dy for Q = 3 GeV, E = 200 GeV, Au–Au collisions producing J/Ψ with $\lambda = 0$.

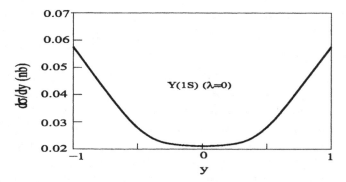

Fig. 10.9.   dσ/dy for Q = 10 GeV, E = 200 GeV, Cu–Cu collisions producing $\Upsilon(1S)$ with $\lambda = 0$.

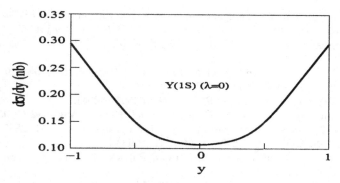

Fig. 10.10. $d\sigma/dy$ for $Q = 10\,\text{GeV}$, $E = 200\,\text{GeV}$, Au–Au collisions producing $\Upsilon(1S)$ with $\lambda = 0$.

This preliminary study can help plan for future studies of the creation of the QGP via A–A RHIC.

6. Problems

1a. From Eqs. (4) and (5), for $y = z = 0$ and $x = 1/b$, what is $B_W(\vec{x})$?

1b. For $y = 0$, $x = 1/b$, and $z = 0.1$ fm, what is $B_W(\vec{x})$?

2. Using the observed Dark Energy density, $\rho_\Lambda^{observed} \simeq (2.3 \times 10^{-3}\ \text{eV})^4$, what is the cosmological constant, Λ?

3. For meson production in p–p collisions, using Eq. (38), for $x = 1$, $m = 1.5\,\text{GeV}$, and energy $E = 200\,\text{GeV}$, what are p_z and $y(x)$?

References

[1] Y. Aoki, Z. Fodor, S.D. Katz and K.K. Szabo, *Phys. Lett.* **B 643**, 46 (2006).

[2] A. Bessa, E.S. Fraga and B.W. Mintz, *Phys. Rev. D* **79**, 034012 (2009).

[3] Anyi Li, Andrei Alexandru and Keh-Fei Liu, *Phys. Rev. D* **84**, 071503 (2011).

[4] Leonard S. Kisslinger, *Phys. Rev. D* **68**, 043516 (2003).

[5] Scott Dodelson, William H. Kinney and Edward W. Kolb, *Phys. Rev. D* **56**, 3207 (1997); William H. Kinney, *Phys. Rev. D* **58**, 123506 (1998); Marc Kamionkowski and Arthur Kosowsky, *Phys. Rev. D* **57**, 685 (1998).

[6] Alexander G. Tevzadze, Leonard Kisslinger, Axel Brandenburg and Tina Kahniashvili, *Astrophys. J.* **759**, 54 (2012).

[7] Dieter Biskamp, *Magnetodynamic Turbulence*, Cambridge University Press, Cambridge (2003).

[8] K. Dolag *et al.*, *Astrophys. J.* **727**, L4 (2011).

[9] Albert Einstein, Die Feldgleichungen der Gravitation, *Sitzungsberichte der Preussischen Akademie der Wissenschatten zu Berlin*, 844 (1915).

[10] Lijuan Zhou, Weiing Ma and Leonard S. Kisslinger, *J. Mod. Phys.* **3**, 1172 (2012)

[11] Federico R. Urban and Ariel R. Zhitnitsky, *Phys. Lett. B* **688**, 9 (2010); *Phys. Rev. D* **80**, 063001 (2009).

[12] Max Tegmark, *et al.*, *Phys. Rev. D* **69**, 103501 (2004).

[13] Hong Jung and Leonard S. Kisslinger, *Nucl. Phys. A* **586**, 682 (1995).

[14] Mikkel B. Johnson and Leonard S. Kisslinger, *Phys. Rev. D* **57**, 2847 (1998).

[15] Nobuyoshi Ohta, *Phys. Lett. B* **695**, 41 (2011).

[16] Leonard S. Kisslinger, Ming X. Liu and Patrick McGaughey, *Phys. Rev. D* **84**, 114020 (2011).

[17] Leonard S. Kisslinger, *Phys. Rev. D* **79**, 114026 (2009).

[18] The CMS Collaboration, arXix: 1105.4894 (2011); *Phys. Rev. Lett.* **107**, 052302 (2011).

[19] G. Moreno *et al.*, *Phys. Rev. D* **43**, 2815 (1991).

[20] V. Khachatryan *et al.*, *(CMS Collaboration)*, *Phys. Rev. D* **83**, 112004 (2011).

[21] Leonard S. Kisslinger, arXiv:1201.1033/hep-ph; MPLA-D-12-00057 (2012).

[22] Leonard S. Kisslinger and Debasish Das, *Mod. Phys. Lett. A* **28(16)**, 1350007 (2013).

[23] Leonard S. Kisslinger, Ming X. Liu, and Patrick McGaughey, preliminary report.

[24] C. Adler *et al. (STAR Collaboration)*, *Phys. Rev. Lett.* **89**, 202301 (2002).

[25] A.D. Frawley, T. Ullrich and R. Vogt, *Phys. Rep.* **462**, 125 (2008).

[26] Rishi Sharma and Ivan Vogt, arXiv:1203.0329/hep-ph (2012).

Solutions to Problems

Chapter 1:

1. A runner, running along a straight track runs 2 miles in 8 minutes. What was his average speed in meters per second?

1. Solution: 1 mile $= 5,280$ feet. 1 foot $= 0.3048$ m. Thus 1 mile $= 1,609.3$ m.

2a. A constant force is acting on a mass. At the initial time $(t = 0)$ the mass is at rest $(v = 0)$, and after 1 minute it has speed $v = 30$ m/s. What is the acceleration?

2a. Solution: Constant acceleration $a = v/t = (30\,\text{m/s})/(60\,\text{s}) = 0.5\,\text{m/s}^2$.

2b. What is the speed after 2 more minutes (3 minutes after $t = 0$)?

2b. Solution: $v = at = 0.5\,\text{m/s}^2 \times (3 \times 60\,\text{s}) = 90\,\text{m/s}$.

3a. What is the circumference in meters of a circle with radius 25 meters?

3a. Solution: Circumference $= 2\pi R = 2\pi \times 25\,\text{m} = 157$ meters.

3b. If a runner runs around this circumference in 1 minute, what was her average speed in meters per second?

3b. Solution: $v = \text{distance/time} = 157\,\text{m}/60\,\text{s} = 2.62\,\text{m/s}$.

4a. If the runner in Problem 3 runs with constant speed, what is the acceleration?

4a. Solution: acceleration $=$ centripetal $= v^2/R = (2.62)^2/25\,\text{m/s}^2 = 0.275\,\text{m/s}^2$.

4b. At any time, what is the direction of the acceleration?

4b. Solution: Toward the center of the circle.

5. A vector $\vec{A}$ has components $A_x = 12\,\text{m}$, $A_y = -6\,\text{m}$. What is the magnitude and direction of $\vec{A}$? Remember if a right triangle has sides a and b, the hypotenuse $c = \sqrt{a^2 + b^2}$.

5. Solution: Magnitude: $A = \sqrt{12^2 + 6^2} = \sqrt{180} = 13.42\,\text{m}$. Direction: $\theta =$ angle between $\vec{A}$ and x-axis. $\sin\theta = -6/13.42 = -0.45$.

6. A vector $\vec{A}$ is 1 m long and is in the x-direction. A second vector $\vec{B}$ is also 1 m long, but is in the y-direction. What is the magnitude and direction of the vector $\vec{C}$, which is the sum of $\vec{A}$ and $\vec{B}$? That is, $\vec{C} = \vec{A} + \vec{B}$.

6. Solution: $C = \sqrt{1+1} = 1.414\,\text{m}$ at an angle of 45 degrees from the x- to the y-direction.

7a. A lead nucleus consists of 82 protons and 126 neutrons. Its mass is about 208 times the proton mass $m_p = 1.67 \times 10^{-27}\,\text{kg}$. If a lead nucleus is dropped with initial speed zero from a height of 1 km, using the acceleration of gravity $= g$ for an object at the surface of earth, how much time does it take for the nucleus to hit the ground?

7a. Solution: Distance $= a \times t^2/2$, $a = g = 9.8\,\text{m/s}^2$ or $t = \sqrt{2 \times 1000\,\text{m}/9.8\,(\text{m/s}^2)} = 14.3\,\text{s}$.

7b. What is the velocity of the nucleus when it hits the ground?

7b. Solution: Velocity $=$ acceleration $\times$ time $= 9.8\,\text{m/s}^2 \times 14.3\,\text{s} = 140\,\text{m/s}$.

7c. What is the kinetic energy of the nucleus in joules when it hits the ground?

7c. Solution: K.E $= (1/2)\,\text{M}v^2 = (1/2) \times 208 \times 1.67 \times 10^{-27}(140)^2$ $\text{kg}(\text{m/s})^2 = 3.40 \times 10^{-21}\,\text{kg}\,(\text{m/s})^2$. 1 joule $= 1\,\text{kg}\,(\text{m/s})^2$, thus K.E. $= 3.40 \times 10^{-21}$ joule.

8. Noting that 1 joule of energy is related to eV by $1\,\text{J} = 6.24 \times 10^{18}\,\text{eV}$, in Problem 7c what is the energy in eV? If all of this energy is turned into heat, what is the temperature T if the matter were at thermal equilibrium?

8. Solution: 1 joule $= 6.24 \times 10^{18}\,\text{eV}$, so K.E. $= 3.40 \times 6.24 \times 10^{-3}\,\text{eV} = 2.12 \times 10^{-2}\,\text{eV}$.

Chapter 2:

1. Yellow light has a wavelength $\lambda = 5.89 \times 10^{-7}$ m. Using Planck's constant $h = 6.63 \times 10^{-34}$ Js and the speed of light $c = 3 \times 10^8$ m/s, how much energy in units of eV does a photon of yellow light deliver to a metal plate in the photoelectric effect?

1. Solution: $1\,\mathrm{J} = 6.24 \times 10^{18}$ eV, so $h = 4.14 \times 10^{-15}$ eV s. $E = hc/\lambda = 4.14 \times 10^{-15} \times 3 \times 10^8/(5.89 \times 10^{-7})\,\mathrm{eV} = 2.11$ eV.

2. What is the wavelength of light emitted as an electron in the $n = 2$ level of a hydrogen atom falls to the $n = 1$ level?

2. Solution: $E_2 - E_1 = -3.4\,\mathrm{eV} - (-13.6)\,\mathrm{eV} = 10.2$ eV. $\lambda = hc/E = 4.14 \times 10^{-15} \times 3 \times 10^8/10.2$ (eV s m/s)/eV $= 1.22 \times 10^{-7}$ m.

3. The mass of the sun $= 1.99 \times 10^{30}$ kg and the mass of the earth is 5.97×10^{24} kg. Using the average distance from the earth to the sun $= 1.5 \times 10^{11}$ m, what is the force of gravity between earth and the sun?

3. Solution: $F = \mathrm{GmM}/R^2 = 6.67 \times 10^{-11} \times 5.97 \times 10^{24} \times 1.99 \times 10^{30}$ $(\mathrm{m}^3/\mathrm{kgs}^2)\mathrm{kgkg}/\mathrm{m}^2 = 7.92 \times 10^{44}$ kgm/s^2.

4. What is the acceleration of gravity 9000 km above the surface of the earth? Note that the radius of the earth is approximately 6,370 km.

4. Solution: $a_g = GM_e/R^2 = 6.67 \times 10^{-11} \times 5.97 \times 10^{24}/\,(9,000 + 6,370)^2$ $(\mathrm{m}^3/\mathrm{kgs}^2)\mathrm{kg}/(1000\,\mathrm{m})^2 = 1.69$ m/s^2.

5a. An electron-neutrino plus a neutron become a proton and an electron. The mass of a neutron is $m_n c^2 = 939.565$ MeV and a proton is $m_p c^2 = 938.272$ MeV. Assuming that the initial neutrino and n and the final p have zero kinetic energy, what is the energy of the electron?

5a. Solution: E(electron) $= 938.565 - 939.272 = 1.293$ MeV.

5b. Assuming that it is not relativistic, what is the velocity of the electron?

5b. Solution: E(electron) $=$ K.E. $+ m_e c^2 =$ K.E. $+ 0.511$ MeV $= 1.293$ MeV, so K.E. $= 0.782$ MeV $= (1/2)\mathrm{m}_e(v^2/c^2)$ MeV. Thus $v^2 = 2\,(0.782/0.511)\,c^2$, $v = 1.75\,c$. This violates Einstein's Special Theory of Relativity.

6a. Draw the Feynman diagram for an electron interacting (weakly) with a proton to create an electron-neutrino and a neutron. What gauge boson was exchanged?

6b. Draw the feynman diagram for a neutrino scattering from a proton. What weak gauge boson was exchanged?

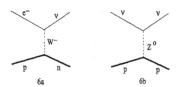

Chapter 3:

1a. A π^+ meson has a lifetime of 2.6×10^{-8} s when it is at rest. An accelerator creates a π^+ meson with a velocity $v = 0.9c$. What is its lifetime in the system of the accelerator?

1a. Solution: Lifetime $(v = 0.9c) = $ lifetime$(v = 0)/\sqrt{1 - (v/c)^2} = 2.6 \times 10^{-8}$ s$/\sqrt{1 - 0.81} = 5.96 \times 10^{-8}$ s.

1b. How far does the pion travel before it decays?

1b. Solution: Distance $= vt = 0.9 \times 3 \times 10^8 \times 5.96 \times 10^{-8}$ (m/s) s $= 17.9$ m.

2. A line in the spectrum of a hydrogen atom has a wavelength $\lambda = 1.3 \times 10^{-7}$ m. If the source of the radiation is moving away from the observer with a speed $v = c/2$, what wavelength is observed?

2. Solution: For the source moving away from the observer $\lambda = \lambda_0 \sqrt{\frac{1+v_s/c}{1-v_s/c}}$ so $\lambda(observed) = 1.3 \times 10^{-7} \times \sqrt{1.5/0.5}$ m $= 2.25 \times 10^{-7}$ m.

3. If a galaxy with a redshift $z = 5$ is emitting the radiation in Problem 2 ($\lambda = 1.3 \times 10^{-7}$ m if the atom is at rest), what wavelength is observed? See table of v/c vs z.

3. Solution: For $z = 5$, $v/c = 0.946$, $\lambda(observed) = 1.3 \times 10^{-7} \times \sqrt{1.946/0.054}$ m $= 7.8 \times 10^{-7}$ m.

4. A rocketship approaching the earth emits a Sodium line, with wavelength $\lambda = 5.896 \times 10^{-7}$ m. Find the wavelength measured on earth if the rocketship's speed is $0.1c$, $0.4c$, and $0.8c$.

4. Solution: Use $\lambda = 5.896 \times 10^{-7} \times \sqrt{(1+(v/c)^2)/(1-(v/c)^2)}$ m, and $\sqrt{(1+(0.1)^2)/(1-(0.1)^2)} = 1.01$, $\sqrt{(1+(0.4)^2)/(1-(0.4)^2)} = 1.175$, $\sqrt{(1+(0.8)^2)/(1-(0.8)^2)} = 2.13$. Therefore $\lambda = 5.96 \times 10^{-7}$ m, 6.93×10^{-7} m, 12.56×10^{-7} m for the three speeds.

Chapter 4:

1. As was derived in Section 2, the speed of the earth moving in a circle with radius $R = 1.5 \times 10^{11}$ m is $v = 3 \times 10^4$ m/s. Using the mass of the earth $= 5.97 \times 10^{24}$ kg and equating centripetal force to gravitational force, find the gravity force between the sun and earth.

1. Solution: Centripetal acceleration $= v^2/R$, Centripetal force $= F_c = mv^2/R$. Thus, gravitational force $= F_g = F_c = 5.97 \times 10^{24} \times (3 \times 10^4)^2/(1.5 \times 10^{11})$ kgm/s^2 $= -3.58 \times 10^{22}$ kgm/s^2.

2a. Our sun is 26,000 ly from the center of our Milky Way galaxy. What is this distance in meters?

2a. Solution: $1\,\mathrm{ly} = 9.46 \times 10^{15}$ m. Distance $= R_s = 2.6 \times 10^4 \times 9.46 \times 10^{15}$ m $= 2.46 \times 10^{20}$ m.

2b. If our sun is rotating about the center of our galaxy at 125 km/s, assuming that all the mass is in a very small volume at the center, what is the mass of our galaxy?

2b. Solution: $a_c = a_g$ or $v^2/R_s = GM_{galaxy}/R_s^2$, $G = 6.67 \times 10^{-11}$ m^3/(kg s^2). Thus $M_{galaxy} = v^2 \times R_s/G = (125 \times 1000\,\mathrm{m/s})^2 \times 2.46 \times 10^{20}$ m/$(6.67 \times 10^{-11}$ m^3/(kg s$^2)) = 5.76 \times 10^{40}$ kg.

3. If during the first ten seconds after the collapse of a star neutrinos are emitted with a momentum of $p = 2 \times 10^{36}$ g km/s, what is the velocity of the neutron star, assuming that it has the mass of the sun?

3. Solution: Using $1\,\mathrm{g} = 10^{-3}$ kg. $p = mv$ or $v = p/M_s = (2 \times 10^{36} \times 10^{-3}$ kg km/s$)/(1.9 \times 10^{30}$ kg$) = 1.05 \times 10^3$ km/s $= 1{,}050$ km/s $=$ a high-luminosity (large T) pulsar velocity.

4. During the ten seconds after the collapse of a massive star, with the neutrinosphere consisting mainly of neutrons, give a reaction that could have produced an anti-muon-neutrino.

4. Solution: $n + n \rightarrow n + p + \bar{\mu}^- + \bar{\nu}_\mu$.

5. As discussed above, matter in the universe about 400,000 years after the Big Bang consisted mainly of hydrogen H and helium He^4. Give a possible set of nuclear reactions that could produce the lead atomic nucleus, Pb^{208}, with 82 protons and 126 neutrons.

5. Solution: Note for $Z > 20$, $N > Z$ for stable nuclei. We use nuclei with (Z, N), $A = Z + N$, having at least a 20% probability for that Z. The nuclei we use have (Z): He(2), Be(4), O(8), S(16), Ni(28), Pd(46), Hg(80), Pb(82). Our reactions: (1) $He^4 + He^4 \to$ Be^8; (2) $Be^8 + Be^8 \to O^{16}$; (3) $O^{16} + O^{16} \to S^{32}$; (4) $S^{32} + S^{32} \to$ $Ni^{58} + 4p$; (5) $Ni^{58} + Ni^{58} \to Pd^{106} + 10p$; (6) $Pd^{106} + Pd^{106} \to$ $Hg^{200} + 12p$; (7) $O^{16} + Hg^{200} \to Pb^{208} + He^4 + 4p$.

Chapter 5:

1. The operator for the x-component of momentum is $\hat{p}_x = -i(h/2\pi)\frac{d}{dx}$. Prove that the function $\langle x|p_x \rangle = e^{ik_x x}$ where $k_x = p_x/\hbar$ is an eigenfunction of the $\hat{p}_x$ quantum operator.

1. Solution: $\hat{p}_x = \frac{\hbar}{i}\frac{d}{dx}$. $\frac{d}{dx}e^{ik_x x} = ik_x e^{ik_x x}$. Therefore $\hat{p}_x e^{ik_x x} = \frac{\hbar}{i}ip_x/\hbar e^{ik_x x} = p_x e^{ik_x x}$.

2. Find the expectation value of $\hat{p}_x$ in the state $|p_x\rangle$.

2. Solution: The expectation value of $\hat{p}_x$ in the state $|p_x\rangle$ is $\langle p_x|\hat{p}_x|p_x\rangle = p_x\langle p_x|p_x\rangle = p_x$.

3a. Starting from $\nu_a = U\nu_\alpha$, with a, α the neutrino flavor, mass, as 1×3 column vectors, show that $\nu_e = \sum_{j=1}^{3} U_{1j}\nu_j$.

3a. Solution: ν_e is in position 1 in the neutrino flavor column matrix, so $\nu_e =$ the first row in the product of the U matrix and the neutrino mass column matrix. Therefore $\nu_e = \sum_{j=1}^{3} U_{1j}\nu_j = U_{11}\nu_1 + U_{12}\nu_2 + U_{13}\nu_3$.

3b. Using the formula in Problem 3a and the 3×3 matrix U given in the text, find ν_e in terms of ν_1, ν_2, ν_3 with the parameters c_{12}, etc.

3b. Solution: $\nu_e = c_{12}c_{13}\nu_1 + s_{12}c_{13}\nu_2 + s_{13}e^{-i\delta_{CP}}\nu_3$.

4. The density of a neutron star is about $3 \times 10^{17}\,kg/m^3$. What would your mass be if you were made of neutron star matter? How much would you weigh on the surface of the earth?

4. Solution: A rough estimate of a typical adult human volume is 1.8 m high and an area of 25 cm × 25 cm, giving a volume

of 0.1125 m³. Since mass = density × volume, mass = 3×10^{17} kg/m³ × 0.1125 m³. Therefore mass = 3.375×10^{16} kg.

Weight = mg = $3.375 \times 10^{16} \times 9.8$ kgm/s² = 3.31×10^{17} kgm/s².

5. A 10 km radius neutron star pulsar is spinning 1000 times a second. What is the speed of a particle on the surface of the pulsar? Compare it to the speed of light.

5. Solution: Distance particle travels is $2\pi R$ = 62.8 km. Speed = distance/time = $62.8 \times 10^3/0.001$ m/s = 6.28×10^7 m/s compared to $c = 3 \times 10^8$ m/s.

Chapter 6:

1a. The F' frame is moving with velocity $v = 0.1\,c$ in the x-direction. If $x = y = z = 0$ and $t = 0$, what are x', y', z', and t'?

1a. Solution: $x' = x = y' = y = z' = z = t' = t = 0$.

1b. Same as 1a. except $t = 1$ s ($x = y = z = 0$). What are x', y', z', t'?

1b. Solution: $x' = (x - ut)/\sqrt{1 - u^2/c^2} = (-0.1 \times 3.0 \times 10^8$ m/s × 1 s)/$\sqrt{1 - 0.01}$ = -3.015×10^7 m. Note $c = 3.0 \times 10^8$ m/s.

2. A rocket 25 m long is moving with speed $v = 0.9c$ with respect to the ground, and a clock in the rocket says $\Delta t' = 30$ seconds, what does an observer on the ground measure as the length of the rocket and the time interval?

2. Solution: $L = 25$ m × $\sqrt{1 - 0.9^2c^2/c^2} = 10.9$ m.

3. A car in the rocket of Problem 2 is moving with speed $v = 0.9c$ with respect to the rocket. What does an observer on the ground measure as the speed of the car?

3. Solution: $v = (0.9 + 0.9)c/(1 + 0.9 \times 0.9) = 0.9945 \times c = 2.983 \times 10^8$ m/s.

4. With the four-velocity U^α and p^α defined in Eqs. (23) and (24), with $\alpha = 0, 1, 2, 3$, what is the scalar product of these two four vectors, $U \cdot p$?

4. Solution: $U \cdot p = U^1 p^1 - U^0 p^0 = \dfrac{mv^2}{1 - (v/c)^2} - \dfrac{\sqrt{p^2c^2 + m^2c^4}}{\sqrt{1 - (v/c)^2}}$.

Chapter 7:

1a. If an object has mass = 1 kg and is moving with speed $v = 0.9c$, what is its momentum in nonrelativistic theory? What is its momentum in the Special Theory of Relativity?

1a. Solution: $p(\text{nonrelativistic}) = mv = 0.9 \times 3.0 \times 10^8 \, (\text{kg m/s}) = 2.7 \times 10^8 \, \text{kg m/s}$.
$p(\text{relativistic}) = mv/\sqrt{1 - v^2/c^2} = 2.7 \times 10^8 \, (\text{kg m/s})/0.4359 = 6.16 \times 10^8 \, \text{kg m/s}$.

1b. What is the relativistic energy (including mass energy as in Einstein's equation for E)?

1b. Solution: $E = \sqrt{p^2/c^2 + m^2} \, c^2 = \sqrt{(6.16 \times 10^8/3.0 \times 10^8)^2 + 1} \times (3.0 \times 10^8)^2 \, \text{kg(m/s)}^2 = 2.055 \times 10^{17} \, \text{kg(m/s)}^2$.

2. What is the Schwarzschild radius for a mass 300 times the mass of the sun?

2. Solution: $R_S = 2GM/c^2 = 2(6.67 \times 10^{-11} \, \text{m}^3/(\text{kg s}^2)) \, 300 \times 1.9 \times 10^{30} \, \text{kg}/(3.0 \times 10^8 \, \text{m/s})^2 = 8.45 \times 10^3 \, \text{m} = 8.45 \, \text{km}$. With $M = $ sun mass, $R_S = 8.45/300 \, \text{km} = 28.2 \, \text{m}$. This is much, much smaller than the radius of the sun, so (as we know) the sun is not a black hole.

3a. Find the ratio of the radius of the universe, $R(t)$, for $t = 10^{11} \, \text{s}$ and $t = 10^{10} \, \text{s}$, in a radiation dominated universe.

3a. Solution: $R(t = 10^{11} \, \text{s})/R(t = 10^{10} \, \text{s}) = \sqrt{10^{11}/10^{10}} = \sqrt{10} = 3.16$.

3b. Same problem as Problem 3a for $R(t = 10^{13} \, \text{s})/R(t = 10^{12} \, \text{s})$, in a matter dominated universe.

3b. Solution: $R(t = 10^{13} \, \text{s})/R(t = 10^{12} \, \text{s}) = (10^{13}/10^{12})^{2/3} = 10^{2/3} = 4.64$.

4. Using the approximate Kolb–Turner formula for $T(t)$, what is the temperature of the universe at time $t = 10^{-10} \, \text{s}$? Compare this to the mass energy of the Higgs.

4. Solution: $T(t)(= kT(t)) = 1 \, \text{MeV}/\sqrt{t} = 1/\sqrt{10^{-10}} \, \text{MeV} = 10^5 \, \text{MeV} = 100 \, \text{GeV}$. Mass energy of the Higgs $= 125 \, \text{GeV} \simeq T(10^{-10} \, \text{sec})$, which will be discussed in Chapter 9.

Chapter 8:

1. Starting from the definition of $\bar{E}$, Eq. (15), carry out the derivation of $\bar{E}$, Eq. (18).

1. Solution: From Eq. (15), $\bar{E} = kT \sum_{n=0}^{\infty} n\nu e^{-na}/\sum_{n=0}^{\infty} e^{-na}$. From Eq. (17), $2\sum_{n=0}^{\infty} e^{-na} = 1/(1 - e^{-a})$. Note $(d/da)\ln(f(a)) =$

$((d/da)f(a))/f(a), (d/da)e^{-na} = -ne^{-na}$. Thus $(d/da)(\ln\sum_{n=0}^{\infty}e^{-na}) =$ $-\sum_{n=0}^{\infty}ne^{-na}/\sum_{n=0}^{\infty}e^{-na} = -\bar{E}/(kTa)$ (Eq. (15)). Also, $((d/da)(1-e^{-a})^{-1})/(1-e^{-a})^{-1} = -e^{-a}(1-e^{-a})^{-1} = -e^{-a}/$ $(1-e^{-a})$. Therefore from Eq. (16), using $a = h\nu/kT$, $\bar{E} = h\nu e^{-a}/$ $(1-e^{-a}) = h\nu/(e^{h\nu/kT}-1)$, which is Eq. (18).

2. Prove that for $h\nu \ll kT$ the Planck spectrum for $\rho(\nu)$ is the same as the Rayleigh–Jeans spectrum.

2. Solution: For $x \ll 1$, $e^x = 1 + x + x^2/2! + \cdots$, therefore for $h\nu \ll kT$, $e^{h\nu/kT} - 1 \simeq h\nu/kT$, so $\rho^{Planck}(\nu) = (8\pi\nu^2/c^3)$ $(h\nu/(e^{h\nu/kT} - 1)) \simeq (8\pi\nu^2/c^3)(h\nu/(h\nu/kT)) = (8\pi\nu^2/c^3)kT = \rho_{R-J}(\nu)$.

3a. Prove that $x(t) = x_o\cos(bt)$ is a solution to the harmonic oscillator equation $\ddot{x}(t) = -b^2x(t)$.

3a. Solution: Using $d^2/dt^2 = (d/dt)(d/dt)$ with x_o a constant, $(d^2/dt^2)(x_o\cos(bt)) = (d/dt)(-bx_o\sin(bt)) = -b^2x_o\cos(bt)$.

3b. Prove that $x(t) = x_oe^{bt}$ is a solution to the harmonic oscillator equation $\ddot{x}(t) = +b^2x(t)$.

3b. Solution: Note that $de^{bt}/dt = be^{bt}$. $(d^2/dt^2)(x_oe^{bt}) = (d/dt)$ $(bx_oe^{bt}) = b^2x_oe^{bt}$.

4. Using Eq. (21), and $a_0^T = 1$, $a_1^T = 0.25$, $a_2^T = 0.1$, plot $T(\hat{n})/T_0$ for $\cos(\theta) = 1$ to 0. I.e., $\theta = 0$ to 90 degrees.

4. Solution: From Eq. (21), $T(\hat{n})/T_0 = 1 + \sqrt{1/(4\pi)}P_0(\cos\theta) +$ $0.25\sqrt{3/(4\pi)}P_1(\cos\theta) + 0.1\sqrt{5/(4\pi)}P_2(\cos\theta) = 1 + 0.282 +$ $0.122P_1(\cos\theta) + 0.063P_2(\cos\theta) = 1 + 0.282 + 0.122\cos\theta + 0.063$ $(1 - (\cos\theta)^2) = 1.345 + 0.122\cos\theta - 0.063(\cos\theta)^2$.

$T(\hat{n})/T_0$ as a function of θ is shown in the figure below:

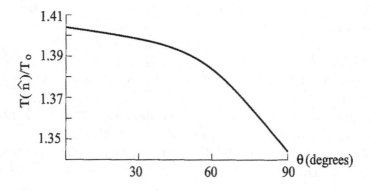

Chapter 9:

1. Prove that the function $\langle x|p_x\rangle = e^{ik_x x}$ where $k_x = p_x/\hbar$ is an eigenfunction of the p_x quantum operator.

1. Solution: $p_x^{op} = \frac{\hbar}{i}\frac{d}{dx} \cdot \frac{d}{dx}e^{ik_x x} = ik_x e^{ik_x x}$, as we have seen. Therefore $p_x^{op}e^{ik_x x} = \frac{\hbar}{i}ip_x/\hbar e^{ik_x x} = p_x e^{ik_x x}$.

2. Prove that $\phi(x) = ae^{ix}$, where a is a constant is a solution to the EOM of Eq. (17).

2. Solution: Eq. (17) is $2\frac{d^2}{dx^2}\phi + 2\phi = 0$. As we have seen, $\frac{d}{dx}e^{ix} = ie^{ix}$. Therefore $\frac{d^2}{dx^2}ae^{ix} = a\frac{d}{dx}\frac{d}{dx}e^{ix} = ai\frac{d}{dx}e^{ix} = -ae^{ix}$, or $2\frac{d^2}{dx^2}ae^{ix} + 2ae^{ix} = 0$, and $\phi(x) = ae^{ix}$ is a solution to the EOM of Eq. (17).

3. Using Eqs. (37), (38), (39), (40) derive Friedmann's equation, Eq. (41).

3. Solution: The 00 component of Eq. (38) is $\mathcal{R}_{00} - \mathcal{R}/2 = 8\pi GT_{00}$, giving $-3\frac{\ddot{R}}{R} + (6/2)(\frac{\ddot{R}}{R} + \frac{(\dot{R})^2}{R^2}) = 8\pi G\rho$ or $\frac{(\dot{R})^2}{R^2} = 8\pi G\rho/3$, which is called Eq. (a).

   The ii component of Eq. (38) is $\mathcal{R}_{ii} - \mathcal{R}/2 = 8\pi GT_{ii}$, giving $-(\frac{\ddot{R}}{R} + 2\frac{(\dot{R})^2}{R^2}) + (6/2)(\frac{\ddot{R}}{R} + \frac{(\dot{R})^2}{R^2}) = 8\pi G(-p)$ or $2\frac{\ddot{R}}{R} + \frac{(\dot{R})^2}{R^2} = -8\pi Gp$, called Eq. (b).

   Subtracting Eq. (a) from Eq. (b), one obtains Friedmann's equation $\frac{\ddot{R}(t)}{R(t)} = -\frac{4\pi G}{3}(\rho + 3p)$, which we used with the cosmological constant to study inflation.

Chapter 10:

1a. From Eqs. (4), (5), for $y = z = 0$ and $x = 1/b$, what is $B_W(\vec{x})$?

1a. Solution: $B_W(x, y, z) = B_W e^{-b^2(x^2+y^2)}e^{-M_n^2 z^2}$. $B_W \simeq 10^{17}$ gauss. Therefore $B(1/b, 0, 0) \simeq 10^{17}e^{-1} = 3.68 \times 10^{16}$ gauss.

1b. For $y = 0$, $x = 1/b$, and $z = 0.1$ fm, what is $B_W(\vec{x})$?

1b. Solution: $hc/M_n \equiv 1/M_n = 6.582 \times 10^{-22}$ MeVs $\times 3 \times 10^8$ (m/s)/939 MeV $\simeq 2 \times 10^{-16}$ m $= 0.2$ fm. Therefore $M_n = 5/$fm and $B(1/b, 0, 0.1\,\text{fm}) \simeq 10^{17}e^{-1}e^{-0.25} = 2.87 \times 10^{16}$ gauss.

2. Using the observed Dark Energy density, $\rho_\Lambda^{observed} \simeq (2.3 \times 10^{-3}\,\text{eV})^4$, what is the cosmological constant, Λ?

2. Solution: $\Lambda = 8\pi G(2.3 \times 10^{-3})^4$, with $G = 6.67 \times 10^{-11}$ m^3/(kg s^2).

3. For meson production in p–p collisions, using Eq. (38), for $x = 1$, $m = 1.5\,\text{GeV}$, and energy $E = 200\,\text{GeV}$, what are p_z and $y(x)$?

3. Solution: Using $p_z = \frac{\sqrt{s}}{2}(x - \frac{a}{x})$, with $\sqrt{s} = E = 200\,\text{GeV}$, $a = 4m^2/s = 4(1.5/200)^2 = 0.000225$, and $x = 1$,

$$p_z \simeq \frac{\sqrt{s}}{2} = 100\,\text{GeV}.$$

$y(x) = \frac{1}{2}\ln(\frac{300}{100}) = 0.239.$

Appendix: Vector Calculus
and Maxwell's Equations

(1) Vector Calculus:

Partial derivative: For $f = f(x)$, a derivative with respect to x is $\frac{d}{dx}f(x) = \frac{f(x+\Delta x)-f(x)}{\Delta x}|_{\Delta x \to 0}$. For $f = f(x,y,z)$, a partial derivative with respect to x is $\partial_x f(x,y,z) = \frac{f(x+\Delta x,y,z)-f(x,y,z)}{\Delta x}|_{\Delta x \to 0}$.

Recall that a vector, $\vec{V}$, has three components, $V_x \equiv V_1$, $V_y \equiv V_2$, $V_z \equiv V_3$.

Scalar Product of two vectors: $\vec{A} \cdot \vec{B} = \sum_{i=1}^{3} A_i B_i$.

Vector Product of two vectors: $(\vec{A} \times \vec{B})_i = A_j B_k - A_k B_j$, with $(i,j,k) = (1,2,3)$, $(2,3,1)$, or $(3,1,2)$ for $i = 1, 2$, or 3.

$(\nabla f)_i \equiv \partial_i f$, with $\partial_i \equiv \frac{\partial}{\partial x_i}$.

$\nabla \cdot \vec{E} \equiv \sum_{i=1}^{3} \partial_i E_i$.

$(\nabla \times \vec{B})_i \equiv \partial_j B_k - \partial_k B_j$ with $(i,j,k) = (1,2,3)$, $(2,3,1)$, or $(3,1,2)$.

(2) Maxwell's Equations and Electromagnetic Field Creation:

In the years 1871–73 James Clerk Maxwell derived the four basic equations of electromagnetism, Maxwell's equations, which he published in a textbook "Treatise on Electricity and Magnetism" (1873). We review Maxwell's equations, which are needed to derive the magnetic fields from the EWPT and QCDPT.

The nonrelativistic Maxwell equations for the magnetic and electric fields ($\vec{B}$ and $\vec{E}$) with $\rho \equiv$ electric charge density and $\vec{J} = q\vec{v}$ the electric current for a particle with charge q and velocity $\vec{v}$, are

$$\nabla \cdot \vec{E} = 4\pi\rho$$

$$\nabla \cdot \vec{B} = 0$$

$$\nabla \times \vec{E} + \frac{1}{c}\frac{\partial \vec{B}}{\partial t} = 0 \tag{1}$$

$$\nabla \times \vec{B} = \frac{4\pi}{c}\vec{J} + \frac{1}{c}\frac{\partial \vec{E}}{\partial t}.$$

The relativistic form of Maxwell's equations are usually given in terms of the scalar Φ and vector $\vec{A}$ potentials, with $\vec{B} = \nabla \times \vec{A}$, $\vec{E} = -\nabla\Phi - \frac{1}{c}\frac{\partial \vec{A}}{\partial t}$. With the notation for the four-vector potential and current $A^0 = \Phi$ and $J^0 = \rho$,

$$\partial^2 A^\nu = \frac{4\pi}{c}J^\nu, \tag{2}$$

and $\nu = 0, 1, 2, 3$. Note that $\partial^2 \equiv \sum_{\nu=0}^{3} \partial_\nu \partial^\nu$.

Equation (2) is used to derive the equation for the magnetic field during the EWPT, which we give without detailed derivation.

The ν component of the B field, B_ν is found from

$$\partial^2 B_\nu - \partial_\nu \partial^\mu B_\mu + |\Phi_H|^2 (g'\sqrt{g^2 + g'^2})A_\nu^{EM} = 0 \tag{3}$$

where $g' = 0.343$, $g = 0.646$ and A^{EM} is the electromagnetic field given by the charged W^+, W^- fields. From symmetry, when the bubbles collide in the z-direction, the B field is a B^ϕ field, in circles around the z direction, as shown in the figure below.

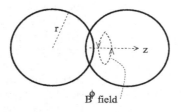

B^ϕ field

The solution for the B field created via EWPT bubble collisions is given in Chapter 9.

The derivation of the B field created via QCDPT bubble collisions is quite different. See Chapter 10.

Index

Printed in the United States
By Bookmasters